A Mathematical
Nature Walk

JOHN A. ADAM

A Mathematical
Nature Walk

PRINCETON UNIVERSITY PRESS

PRINCETON AND OXFORD

Copyright © 2009 by Princeton University Press
Published by Princeton University Press, 41 William
Street, Princeton, New Jersey 08540
In the United Kingdom: Princeton University Press,
6 Oxford Street, Woodstock, Oxfordshire OX20 1TW
press.princeton.edu

Fifth printing, and first paperback printing, 2011
Paperback ISBN 978-0-691-15265-3

The Library of Congress has cataloged the cloth edition
of this book as follows

Adam, John A.
  A mathematical nature walk / John A. Adam.
    p. cm.
  Includes bibliographical references and index.
  ISBN 978-0-691-12895-5 (hardcover : alk. paper)
1. Mathematics in nature—Miscellanea.
2. Mathematical analysis—Miscellanea.   I. Title.
  QA99.A23 2009
  510—dc22   2008044828

British Library Cataloging-in-Publication Data
is available

This book has been composed in Minion
Printed on acid-free paper. ∞

Printed in the United States of America

10   9   8   7

For Susan

When through the woods and forest glades
  I wander
And hear the birds sing sweetly in the trees,
When I look down from lofty mountain grandeur
And hear the brook and feel the gentle breeze,
Then sings my soul . . .
                    —*Carl Gustav Boberg (1859–1940)*

Contents

Preface

As the reader will realize from the title of this book, if nothing else, I like to walk a lot! And I see many fascinating phenomena in the sky, on the water, and everywhere I go. My wife, Susan, and I love to visit state and national parks, and some of the questions in this little volume reflect that. So perhaps I should not have been surprised one evening, as I was drifting off to sleep, when a thought suddenly appeared in my mind: why not have a "mathematical nature walk" as a motif? Tired though I was, this seemed an eminently sensible idea, and I woke up the next morning quite excited about this way of arranging the book. I was also relieved that I'd remembered the thought at all!

So, for whom is this book written? It is for anyone who, while outside in some capacity, whether gardening, walking, or driving, has wondered, "What *is* that?" What is that splotch of color near the sun? Why is that cloud colored? Why do those waves appear like that? What makes that shadow appear so unusual? There are other questions that may not have come to mind (such as emptying Loch Ness, and combustible haystacks), but may, nevertheless, intrigue and encourage the reader to think hard about a particular topic. In short, the book is written for anyone interested in nature, and who has a willingness to think, question, and encounter a modicum of mathematics along the way. Certainly teachers of science and mathematics should fall into this category! I hope that as a result of reading the book, the reader will gain more of an understanding of both the beauty of nature and some of the physical principles and mathematical structure underlying this beauty. And this essentially answers another question: Why did I write this book at all? I am interested in nature, and ever since I read an introductory account of calculus when I was sixteen years old, I have been fascinated by the application of mathematics to the world of nature. There are always more and more questions to be asked, and nothing, it seems to me, is ever *completely* explained; whether the explanation is understood in broad terms only, or in great detail, there are frequently ever more subtle features of a given phenomenon that require yet further investigation. This is what makes science in general and applied mathematics in particular so fascinating (no matter how elementary the level). The structure of the book is laid out below, but at the outset let me encourage the reader to dip into the book here and there. There is absolutely no necessity to read it in the order the questions appear, though, of course, some are related to one another. That should be clear from the context of the question.

What are the mathematical prerequisites for reading this book? Some of the questions require arithmetic only (such as Q.1–Q.23), and the majority of the remaining questions are accessible to those with a background in precalculus (algebra and trigonometry). The rest of the questions require elementary calculus, and some very simple differential equations, though they need not be ignored by those without this background, for there is much to be gained from the descriptive material surrounding the equations. And how is this book different from the earlier one, *Mathematics in Nature* [*MiN*]? While several of the topics embedded in [*MiN*] have been "unfolded" here to make them more accessible to a wider audience, this book is not intended to be a sequel. My intention in writing that earlier volume was to address, in an increasingly technical manner, a wide variety of naturally occurring patterns. For *A Mathematical Nature Walk*, the idea of posing questions seemed to be a good way to introduce the subject, objectives, and objects of modeling in a rather more low-key and perhaps provocative fashion. Several exercises for the reader, based on the questions and answers, are scattered throughout the book. Neither book was written as a textbook per se, though there is nothing to prevent either one being used as a supplement to an existing course in modeling, for example. Indeed, an instructor's manual has been written for [*MiN*].

In the Introduction, to set the scene for the rest of the book, the philosophy and broad methodology of mathematical modeling are discussed and illustrated with a simple model of a melting snowball. This is followed by some thoughts on *inverse problems* that are in fact much more important in the history of applied mathematics in particular and science in general than is usually realized.

After the Introduction, the book is divided into sections representative of the location of the walks. Some of these are quite long, and others are very short, and they are not necessarily mutually exclusive. The first section, grandiosely entitled *At the beginning*, is really just a collection of general questions designed to challenge the reader's powers of observation, estimation, and physical intuition, and to whet the appetite for the fun to follow. *In the "playground"* is just such fun, from my perspective at least, for it mostly consists of romping around with "mere" numbers, so to speak, in the spirit of Fermi problems, after the legendary physicist Enrico Fermi, who took great pleasure in creating and solving them (many such problems are in the book by

Weinstein and Adam). In this playground we get our feet wet, or sandy, or both. Then follow in rapid succession sections entitled *In the garden*, *In the neighborhood*, and *In the shadows*. Far from being sinister, this latter topic is just about that: shadows. But shadows are mysterious things, without substance and yet without a doubt real, and so frequently we just take their existence for granted. And as mentioned above, some topics make their appearance in other sections, and shadows are no exception.

In the sky, the largest section in the book, is next. But no sooner are our eyes upward facing, than perhaps a passing bird or the sight of a nest high in tree might cause us to focus our attention on a new but single topic: birds' eggs. The section *In the nest* provides a look at how and why several complimentary mathematical models have been developed in the past using, for the most part fairly elementary, but none the less fascinating techniques. *In (or on) the water* implies an encounter with waves and the ducks and ships that produce some of them, but there is more to it than that. This is followed by *In the forest*, *In the national park,* and *In the night sky*. The latter contains a question about modeling a total solar eclipse; momentarily then, the sky is as dark as at night, though clearly the phenomenon is a daylight one. The final section of questions is not surprisingly called *In the end*, and is so-called for more than just the obvious reason.

Many drawings and photographs are included to illustrate the questions and their explanations. Color photographs are grouped together in a special section in the middle of the book, organized by question number. A bracketed reference to a color plate is provided in the relevant questions to direct the reader to this section.

There are four appendixes, the first being a very short glossary of a few of the mathematical terms and functions referred to in the book. I have included only a few such functions: the hyperbolic tangent function, the trigonometric inverse sine and cosine functions, and perhaps a surprising guest, the Bessel function of the first kind of order zero. Ordinary trigonometric functions make many appearances, of course, but I did not think it necessary to include them since they are so common. Appendix two summarizes the "answers" (some of them necessarily a little imprecise) to the questions posed in *At the beginning*. The third appendix discusses Newton's law of cooling in the context of my penchant for coffee, especially while writing this book. It may seem somewhat out of place, but my reason for its inclusion follows the development of one such simple model for the growth of a "tree tumor" or burl (question 83). Finally, appendix four is a compilation of many of the patterns to be noticed in the natural world about us. While some are discussed in some detail in this book, the list is certainly not exhaustive, and the reader is encouraged to be on the look out for them, and many others!

Acknowledgments

I would like to thank Heather Renyck for permission to use her photographs in questions 68 and 75; Dave Lynch for allowing me to use his photograph of a ship wake in question 78, and also for his careful reading of the manuscript; Wally Berg (www.bergadventures.com) for permission to use the shadow of Everest at sunrise photograph in question 85 (and to Bill Livingston for putting me in touch with him!); Michael Vollmer for allowing me to quote extensively from his paper with K.-P. Möllmann, modeling solar eclipses (question 92); and Roger Pinkham for pointing out to me a succinct argument leading to Rayleigh's result, equation (81.2), in question 81.

My chairman, Mark Dorrepaal, has been very generous in arranging my teaching schedule to accommodate, wherever possible, my many speaking and writing engagements. I could not have completed this and other projects without his enlightened support. I am very grateful to Barbara Jeffrey for scanning the scraps of (often) hand-drawn figures so that I could pass them on to Don Emminger, whom I thank for his careful and painstaking work in transforming those scans to what you see in this book today.

I am very pleased to acknowledge the continued support and encouragement from my friend and fellow "EPOD-er," Heather Renyck, geology teacher and nature lover *extraordinaire*. Her zest for the world around her—animal, vegetable, but especially mineral—has amazed me. And in Cathleen Horne I have found a kindred spirit; she loves the confluence of mathematics and nature as much as I do! I thank both of them for the gift of their friendship. My editor, Vickie Kearn, as always, has been a blessing to me, and a source of sage advice and counsel. She just "gets it," and I thank her for helping me to stay on top of the writing, even when other commitments tended to delay my self-imposed deadlines. Her assistant, Anna Pierrehumbert, went through the manuscript in great detail—I have to conclude she actually read it—and suggested many valuable improvements. I am also very grateful for the detailed comments and constructive criticisms from two self-identified reviewers of a previous incarnation of the book: Will Wilson, and Brian Sleeman. Thanks to them, particularly Will, I was encouraged to restructure it significantly and, I hope, for the better. A third reviewer made many helpful suggestions, also resulting, I trust, in a much improved manuscript.

The deepest debt of gratitude I owe is to my wife, Susan, who urged me to think very carefully about the intended audience for this book, and made many valuable suggestions for topics to include (and exclude!). Truly, she is a loving companion who has always had my best interests, both professional and personal, at heart.

A Mathematical
Nature Walk

Introduction

I have long been fascinated by the wonders of nature encountered on a simple walk, even in residential areas, as long as there is a view of the sky, roads lined with trees, and bodies of water within easy walking distance. Such is the case where I live in Norfolk, Virginia, and while most of my nature walks are local in character, I have been fortunate to have had the opportunity to travel farther afield at times, both within the United States and beyond its borders. Many of the topics in this book are consequences of such opportunities. And just as some walks are fairly short, and some are longer—veritable hikes, indeed—so are the chapter sections correspondingly varied.

Needless to say, the questions posed, and answered, are based on *many* walks, sometimes only into the back garden, or across the street, to gain a better view. I doubt that more than a few of these topics could be encompassed in a single walk, no matter how long! Mathematically, some are more involved than others; some are merely short forays into the topic, thinking on paper, as it were. Indeed, the shorter answers are perhaps best seen as "toy" models—merely intended to whet the appetite, to stimulate and encourage the reader to pursue the topic further on another occasion. Indeed, frequently, in my experience, at least, thinking about a question may lead nowhere initially, and it has to be put aside for a time. Later, however, it may yield some fascinating and valuable insights, perhaps of pedagogic value. The reader is encouraged always to take such "answers" and use them as a starting point to develop a fuller appreciation for the topic.

There is one question that should be addressed at the outset in a book like this, however. It is *Can you really use mathematics to describe that?* Regardless of whatever "that" may be, this is a question with which most people who teach the subject can identify, I suspect. Mathematically self-effacing, but otherwise educated people are frequently surprised when, in casual conversation, the subject of mathematical modeling comes up, often, in my experience, in connection with biological problems. Most people are at least vaguely aware of the importance of mathematics in physics, chemistry, or astronomy, for example, perhaps through unfortunate encounters with those subjects in earlier years. However, much less common is the notion that other subjects are also amenable to mathematical modeling in this way, and so part of the motivation for this book has been to address questions of this type by identifying an eclectic collection of "how to model [put your favorite topic here]" chapters. The unifying themes are these: *mathematical modeling* and *nature*.

What, then, *is* a mathematical model? Under what kinds of assumptions can we formulate such a model, and will it be realistic? Let us try and illustrate the answer in two stages; first, by drawing on a simple model discussed elsewhere (Adam 2006) in answer to the question: *Half the mass of a snowball melts in an hour. How long will it take for the remainder to melt?* Following that, a more general account of modeling is presented. This type of problem is often posed in first-year calculus textbooks, and as such requires only a little basic mathematical material, e.g., the chain rule and elementary integration. However, it's what we do with all this that makes it an interesting and informative exercise in mathematical modeling. There are several reasonable assumptions that can be made in order to formulate a model of snowball melting; however, unjustifiable assumptions are also a possibility! The reader may consider some or all of these to be in the latter category, but ultimately the test of a model is how well it fits known data and predicts new phenomena. The model here is less ambitious (and not a particularly good one either), for we merely wish to illustrate how one might approach the problem. It can lead to a good discussion in the classroom setting, especially during the winter. Some plausible assumptions might be as follows.

(i) Assume the snowball is a sphere of radius $r(t)$ at all times. This is almost certainly never the case, but the question becomes one of simplicity: is the snowball roughly spherical initially? Subsequently? Is there likely to be preferential warming and melting on one side even if it starts life as a sphere? The answer to this last question is yes: preferential melting will probably occur in the direction of direct sunlight unless the snowball is in the shade or the sky is uniformly overcast. If we can make this assumption, then the resulting surface area and volume considerations involve only the one spatial variable r.

(ii) Assume that the density of the snow/ice mixture is constant thoughout the snowball, so there are no differences in "snow-packing." This may be reasonable for small snowballs (i.e., hand-sized ones) but large ones formed by rolling will probably become more densely packed as their weight increases. A major advantage of the constant density assumption is that the mass (and weight) of the snowball is then directly proportional to its volume.

(iii) Assume the mass of the snowball decreases at a rate proportional to its surface area, and only this. This appears to make sense since it is the outside surface of the snowball that is in contact with the warmer air, which induces melting. In other words, the transfer of heat occurs at the surface. This assumption in particular will be examined in the light of the model's prediction. But even if it is a good assumption to make, is the "constant" of proportionality really constant? Might it not depend on the humidity of the air, the angle of incidence and intensity of sunlight, the external temperature, and so on?

(iv) Assume that no external factors change during the "lifetime" of the snowball. This is related to assumption (iii) above, and is probably the weakest of them all. Unless the melting time is very much less than a day it is safe to say that external factors will vary! Obviously, the angle and intensity of sunlight will change over time, and also possibly other factors as noted above.

Let's proceed on the basis of these four assumptions and formulate a model by examining some of their mathematical consequences. We may do so by asking further questions. For example,

(i) What are expressions for the mass, volume, and surface area of the snowball?

(ii) How do we formulate the governing equations? What are the appropriate initial and/or boundary conditions? How do we incorporate the information provided?

(iii) Can we obtain a solution (analytic, approximate, or numerical) of the equations?

(iv) What is the physical interpretation of the solution and does it make sense? That is, is it consistent with the information provided and are the predictions from the model reasonable?

Let $r(t)$ be the radius of the snowball at time t hours after the start of our "experiment," and let the initial radius of the snowball be $r(0) = R$. The surface area of a sphere of radius r is $4\pi r^2$ and its volume is $4\pi r^3/3$. If we denote the uniform density of the snowball by ρ, then the mass of the snowball at any time t is

$$M(t) = \tfrac{4}{3}\pi \rho r^3(t). \tag{I.1}$$

The instantaneous rate of change of the mass of the snowball (the derivative of $M(t)$ with respect to t) is then

$$\frac{dM}{dt} = 4\pi \rho r^2 \frac{dr}{dt}. \tag{I.2}$$

By assumption (iii), dM/dt is proportional to $S(t)$, the surface area at time t:

$$\frac{dM}{dt} = -4\pi r^2 k, \tag{I.3}$$

where k is the (positive) constant of proportionality, the negative sign implying that the mass is *decreasing* with time. By equating the last two expressions it follows that

$$\frac{dr}{dt} = -\frac{k}{\rho} = -\alpha, \text{ say.} \tag{I.4}$$

Note that this implies that according to this model, the radius of the snowball decreases uniformly with time. This means that the radius $r(t)$ is a linear function of t with slope $-\alpha$; since the initial radius is R, we must have

$$r(t) = R - \alpha t = R\left(1 - \frac{t}{t_\mathrm{m}}\right) = 0 \text{ when } t = \frac{R}{\alpha} = t_\mathrm{m}, \tag{I.5}$$

where t_m is the time for the original snowball to melt, which occurs when its radius is zero. However, we do not know the value of α since that information was not provided, but we *are* informed that after one hour, half the snowball has melted, so we have from equation (I.5) that $r(1) = R - \alpha$. A sketch of the linear equation in (I.5) and use of similar triangles (figure I.1) shows that

$$t_\mathrm{m} = \frac{R}{R - r(1)}, \tag{I.6}$$

and furthermore

$$\frac{V(1)}{V(0)} = \frac{1}{2} = \frac{r^3(1)}{R^3}, \tag{I.7}$$

so that

$$r(1) = 2^{-1/3} R \approx 0.79 R.$$

Hence, $t_\mathrm{m} \approx 4.8$ hours, so that according to this model the snowball will take a little less than 4 more hours to melt away completely. This is a rather long time, and certainly the sun's position will have changed during that time

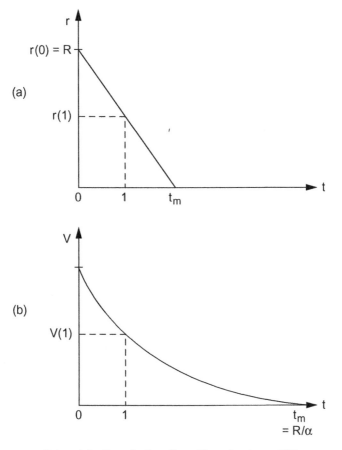

Figure I.1. Snowball radius $r(t)$ and volume $V(t)$

(through an arc of about $70°$), so in retrospect assumption (iv) is not really justified. A further implication of equation (I.5) is that the volume (and also mass of the snowball by assumption (ii)) decreases like a cubic polynomial in t, i.e.,

$$V(t) = V(0)\left(1 - \frac{t}{t_m}\right)^3. \tag{I.8}$$

Note that $V'(t) < 0$ as required, and $V'(t_m) = 0$. Since $V''(t) > 0$, it is clear that the snowball melts more quickly at first, when $|V'|$ is larger, than at later times, as figure I.1(b) attests. I recall being told as a child by my mother that "snow waits around for more" but this model is hardly a proof of that, despite further revelations below! It may be adequate under some circumstances but there are obvious deficiencies given the initial data (which I invented). What other factors have been ignored here? Here are some.

We are all familiar with the fact that the consistency of snow varies depending on whether it is "wet" or "dry"; snowballs are more easily made with the former. Wet snow can be packed more easily and a layer of ice may be formed on the outside. This can in turn cool a thin layer of air around the surface, which will insulate (somewhat) the snowball from the warmer air beyond that. A nice clean snowball, as opposed to one made with dirty snow, may be highly reflective of sunlight (i.e., it has a high *albedo*) and this will reduce the rate of melting further. There are no doubt several other factors missing.

Some other aspects of the model are more readily appreciated if we generalize the original problem by suggesting instead that a fraction β of a snowball melts in h hours. The melting time is then found to be

$$t_{\mathrm{m}} = \frac{h}{1 - \sqrt[3]{1 - \beta}}, \tag{I.9}$$

which depends linearly on h and in a monotonically decreasing manner on β. The dependence on h is not surprising; if a given fraction β melts in half the time, the total melting time is also halved. For a given value of h, the dependence on β is also plausible: the larger the fraction that melts in time h, the shorter the melting time.

Leaving the snowball to melt, we have seen in an elementary way from this example that certain fundamental steps are necessary in developing a mathematical model (see figure I.2): formulating a real world problem in mathematical terms using whatever appropriate simplifying assumptions may be necessary; solving the problem thus posed, or at least extracting sufficient information from it; and finally interpreting the solution in the context of the original problem. Thus, the "art" of good modeling relies on (i) a sound understanding and appreciation of the problem, such as what factors affect the melting rate of the snowball; (ii) a realistic, but not unnecessarily mathematical representation of the important phenomena; (iii) finding useful solutions, preferably quantitative ones, and (iv) interpretation of the mathematical results—yielding insights, predictions, such as when the snowball will melt away completely, and so on. Sometimes the mathematics used can be very simple, as above; indeed, the usefulness of a mathematical model should not be judged by the sophistication of the mathematics, but by its predictive capability, among other factors. Mathematical models are not necessarily "right" (though they may be wrong as a result of ignoring result of ignoring fundamental processes). One model may be better than another in that it has

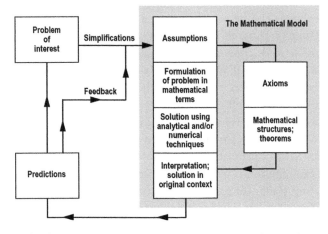

Figure I.2. Steps associated with developing a mathematical model

better explanatory features, or more specific predictions can be made, which are subsequently confirmed, at least to some degree. Sometimes models can be controversial: this is a good thing, for it generates scientific and mathematical discussion. Indeed, I venture to go further and suggest that all mathematical models are flawed to some extent: many by virtue of inappropriate assumptions made in formulating the model, or (which may amount to the same thing) by the omission of certain terms in the governing equations, or even by misinterpretation of the mathematical conclusions in the original context of the problem. Occasionally, models may be incorrect because of errors in the mathematical analysis, even if the underlying assumptions are valid. And paradoxically it can happen that even a less accurate model is preferable to a more mathematically sophisticated one; it was the mathematical statistician John Tukey who stated that "it is better to have an approximate answer to the right question than an exact answer to the wrong one." But another comment must be made about modeling, and mathematical problems in general.

Inverse problems: what is a question to which the answer is . . . ?

> Usually in mathematics you have an equation and you want
> to find a solution. Here you were given a solution and you had
> to find the equation. I liked that.
> —Julia Robinson (as quoted by C. W. Groetsch)

Many problems can be posed in formal terms, such as $A \Longrightarrow ?$, meaning *what does A imply*? Thus, we might ask, *What is 47 times 59?* or *Given a bird's egg, is it possible to describe its shape mathematically?* (see Q.62–Q.67) or, as posed above, *How long will it take a snowball to melt?* Most people encounter only these so-called *direct problems* in their mathematical education. But there is another side to all this, indeed, a whole new universe of potential questions in the form of *What implies A?* Life is full of such *inverse problems*, as they are called, though we may be forgiven for not recognizing it. Given a "spot" on an X-ray, diagnosticians seek to determine what has caused it. An ultrasound examination is used to determine whether the baby in the womb is a girl or a boy. Radar (or sonar) is used to infer "what is out there," or more precisely, what object is causing this particular pattern of electromagnetic (or acoustic) wave reflection. I've long been fascinated by the concept of inverse vs. direct problems. Let me try to explain more precisely what I mean by these descriptors. There are many levels on which we can think about this, the simplest perhaps being with regard to multiplication: as above, let's ask, *What is the product of 47 and 59? 2773*, you instantly reply. Now, find a pair of factors for 9831. It's not quite such an easy task to find the factors 87 and 113 (and there are other possible pairs because 87 is 3 times 29, i.e., the prime factors are 3, 29, and 113, so clearly the decomposition is not unique).

And, incidentally, this illustrates another point: as with many direct problems, answers to inverse problems may not be unique. In a fascinating article on these problems, Joseph Keller introduces this point with three inverse word problems, one of which is *What is a question to which the answer is 9W?* In a classroom setting this invariably produces the standard question, *What is 9 times W?*, dismissed immediately (though correct) as being far too pedestrian! I try to encourage the students to think a little more "outside the box," so we do get questions like *What's your shoe size?* (so perhaps it's a shoebox) and *What route gets you from "A" to "B"?*, and these are quite satisfactory. Keller's question, wonderfully creative (and, in my view, very funny), is, *Do you spell your name with a "V", Herr Wagner?* Herr Wagner's answer is, in fact, *Nein, W*.

Another elementary example, this time from ballistics, is valuable, especially if we can neglect air resistance and the curvature and the rotation of the earth, so the only force to consider is that of gravity. The direct problem is then to determine the range of the projectile (a cannonball, say) given its initial speed and the angle of elevation of the barrel. There is a readily found unique solution to this problem, using Newton's second law of motion: neglecting air resistance, the distance d traveled by a projectile with angle of elevation α ($0 < \alpha < \pi/2$) and initial speed u_0 is given by the formula

$$d = \frac{u_0^2}{g} \sin 2\alpha. \tag{I.10}$$

Therefore, the direct problem (given u_0 and α, find d) has a unique solution. On the other hand, given the range of the cannonball with u_0 fixed, there may be zero, one, or two possible solutions for the inverse problem: if $d < u_0^2/g$ is specified, then these solutions can be seen to exist from the graph of equation (I.10), and they are symmetrically placed about the line $\alpha = \pi/4$, the angle for maximum range d. Analytically, the solutions are

$$\alpha_1 = \frac{1}{2} \arcsin\left(\frac{gd}{u_0^2}\right), \quad \alpha_2 = \frac{\pi}{2} - \frac{1}{2} \arcsin\left(\frac{gd}{u_0^2}\right), \tag{I.11}$$

as is easily verified.

Exercise: Establish the results (I.10) and (I.11).

As pointed out by Groetsch in the book cited below, many of the major historical breakthoughs in science are a result of solving, in essence, an inverse problem. Thus, the curved shadow of the earth on the surface of the moon during a lunar eclipse enabled Plato's student Aristotle (384–322 BC) to infer (among other arguments) that the earth was spherical. This indirect form of reasoning was used, still within an astronomical context, by Newton to derive the inverse-square law of gravitation from Kepler's laws of planetary motion, which were themselves inferred from Tycho Brahe's observations. Furthermore, the young English mathematician John Couch Adams (1819–1892) and the French mathematician Urbain LeVerrier (1811–1877) independently used deviations in the observed position of the outermost of the planets known at the time, Uranus, to infer the existence and position of a perturbing body beyond Uranus: the planet Neptune. When telescopes were turned to the region of sky predicted by the mathematics, the inverse problem and its solution were validated—there was the new planet!

These simple models illustrate much of the fundamental approach I have taken in the "walks" discussed in this book. None are exhaustive or complete; and therefore all are open to improvement. Some are very basic and simplistic (of the "back-of-the-envelope" variety), while others are more sophisticated. Some are direct problems and some are inverse problems (try to determine which). So from my backyard to the Rockies and beyond, allow me to invite you to accompany me on my mathematical nature walks.

References

Adam, J. A. (Paperback edition 2006). *Mathematics in Nature: Modeling Patterns in the Natural World.* Princeton University Press, Princeton, NJ.

Groetsch, C. W. (1999). *Inverse Problems: Activities for Undergraduates.* The Mathematical Association of America, Washington, DC.

Keller, J. B. (1976). "Inverse Problems." *The American Mathematical Monthly* 83, 107–118.

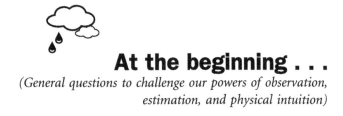

At the beginning . . .

(General questions to challenge our powers of observation, estimation, and physical intuition)

Below are general questions to test our powers of observation, estimation, and physical intuition. I think that they are a valuable way to judge just how much we really *notice* what goes on around us, and they provide a fruitful basis for discussion when introduced, say, in a classroom. The "answers" (some of which are of necessity fuzzy), are found in Appendix 2. So, without further ado, we naturally start—

Q.1:   You are looking at a single bright rainbow (the primary bow). Which color is on the top side of the arch?

Q.2:   How many colors do you think there are in a typical rainbow?

Q.3:   Suppose you see a double rainbow; which color is on the top side of the upper arch (the secondary bow)?

Q.4:   Is the region between the two bows typically darker, brighter, or the same as the surrounding sky?

Q.5:   How about the region below the primary bow?

Q.6:   Have you ever seen anything else closely associated with the primary bow?

Q.7:   On a sunny day, are the edges of your shadow sharp, diffuse, or some combination of both? Explain your answer as precisely as you can.

Q.8:   Over what typical timescale does a small cumulus cloud maintain its original shape (i.e., from when you first looked at it)?

Q.9:   Estimate the size (diameter) of water droplets in (i) a heavy downpour, and (ii) fog.

Q.10:   To the nearest order of magnitude, how many light waves would fit across your fingernail?

Q.11:   How long is a typical sound wave associated with human speech?

Q.12:   What is the (eastward) rotation speed of the earth at the equator? At the poles?

Q.13:   At what latitude is the rotation speed the arithmetic mean of the correct answers to the two parts of the previous question?

Q.14:   How far away is the horizon if you are standing at the beach looking out to sea?

Q.15:   In the middle of the day, you are looking at two similar hills, except that one is more distant than the other. Which one appears a bit darker? (i) the nearer one; (ii) the more distant one; (iii) both appear equally dark, or (iv) take off your sunglasses, silly!

In the "playground"
(just to get our feet wet . . .)

It is the mark of an instructed mind to rest satisfied with the degree
of precision which the nature of the subject permits and not to seek
an exactness where only an approximation of the truth is possible.
—Aristotle, *Nicomachean Ethics*

We have set the scene for our nature walks, so let's get
started.

In this section, we'll just attempt some "applied arithmetic" and a little bit
of algebra. Even if some of the following estimation problems could be calculated exactly (and most cannot), that is not the point here: *most* things
cannot be computed exactly, and frequently all that is necessary is a simple
estimate. By breaking down a problem into simpler subproblems, making
plausible assumptions and using simple arithmetic, we can genuinely arrive at
an order of magnitude "guesstimate" of the answer. Obviously, we may each
make somewhat different assumptions, but it is as unlikely that all of mine
will be *overestimates* as it is that all of yours will be *underestimates*. This being
the case, these differences between our approaches will tend to cancel out,
and, by and large, our final results will probably be in the same ballpark. Hans
Christian von Baeyer in his book *The Fermi Solution* has made the comparison of this process with the results of a series of throws of a coin: it is very
improbable that, say, ten coin tosses will result in a string of ten heads or ten
tails (less than one chance in a thousand, in fact; do you see why?).

Q.16:   Loch Ness—how long to empty it?

Loch Ness has a volume of about 2 cubic miles. How many times would a
single one-gallon bucket need to be filled to empty the loch (and thus expose
the monster)?

This exposes my British upbringing like nothing else, but, hey, why do things simply in the metric system when making it complicated is so easy?* Now one cubic foot of water is approximately 7.5 gallons, so one gallon ≈ 0.1 cubic foot. Since the volume of Loch Ness is approximately $2 \times (5 \times 10^3)^3 \approx 3 \times 10^{11}$ cubic feet, this is equivalent to about 3×10^{12} bucketfuls of water. Now let's assume for simplicity that you can stand at the (decreasing) edge of the loch and just scoop the water into trucks that carry it all away. A gallon of water is pretty heavy to handle repeatedly, but we'll further assume that you have an army of volunteers from the "Capture Nessie Society" who are willing and able to take over when you have muscle fatigue. So let's now assume that each "scoop and empty" operation takes about 10 seconds, so you manage about 300 bucketfuls per hour (allowing some time for a few extra breaths). This is very convenient: in hours, the time it will take is about $(3 \times 10^{12}) \div 300$, or 10^{10} hours! There being about 400 days in a year (rounding up) and about 20 hours in a day (rounding down), that's $10^{10} \div (400 \times 20) \approx 10^6$ years. Nessie will probably be dead by then (and so will we). Hmm . . . Okay guys, back to the drawing board.

---

*All right, I'm ashamed. Who would know the conversion from cubic feet to gallons? Two cubic miles is approximately $2 \times (1.6)^3 \approx 10 \ (\text{km})^3 = 10 \times (10^5)^3 \ \text{cm}^3 = 10^{16} \ \text{cm}^3$, and $10^3 \ \text{cm}^3$ is one liter, so this is about 10^{13} liters or about 3×10^{12} bucketfuls as before.

Q.17:   The Grand Canyon—how long to fill it with sand?

(i) One of Gary Larson's *Far Side* cartoons depicts "Larry" emptying sand out of a dump truck into the Grand Canyon, and filling it up while no one noticed ☺. Estimate how long it might take to actually accomplish this!

(ii) How long would it take Larry, by now flushed with success from Arizona, to demolish Mt. Fuji by carting away the rock and soil in his dump truck?

(i) Obviously we need some information on the physical characteristics of both the above features! According to the official Grand Canyon website, the average depth of the Canyon along its 277-mile length is 4000 ft (the greatest depth being about 6000 ft). Its greatest width is 15 miles, but no average is noted; I shall make a conservative estimate of 5 miles as the average width. What's the size of a typical dump truck? I shall estimate a truck bed size of about 12 ft × 6 ft × 4 ft ≈ 300 cu. ft (this may be an underestimate), so mentally replacing the Grand Canyon by a rectangular box of dimensions 1 mile× 200 miles×5 miles, or 1000 cubic miles. This is approximately $10^3 \times (5 \times 10^3)^3 \approx 10^{14}$ cu. ft, or approximately 3×10^{11} dump truck loads. Assuming Larry, and his family, if necessary, has a turn-around time of one hour to fill the truck with sand or dirt, drive to the Canyon, empty the truck, and return to the site (wherever that is—Japan, maybe (see part (ii) below), in which case we need to extend the time interval somewhat ☺), then if they work all day, this would require about $[(3 \times 10^{11})/24] \approx 10^{10}$ days $\approx 3 \times 10^7$ years!

(ii) From information and photographs of Mt. Fuji, I take it to be a right circular cone 12,000 ft in height and with base diameter about twice that. The volume in cubic feet is therefore about $(\frac{1}{3}) \times 3 \times (10^4)^2 \times 10^4 = 10^{12}$ which corresponds to about 3×10^9 truck loads. At the same rate as in part (i) this would correspond to 10^8 days $\approx 3 \times 10^5$ years. Perhaps Larry should start with something even less ambitious.

Q.18:   **Just how large an area is a million acres?**

> Arizona fires unite, march on homes; flames could engulf 1 million acres before it's over, officials say.
> —Headline on the front page of the *Virginian Pilot*, June 24th, 2002

Answer: 1 million acres. While correct, it adds no new insights to the question. It must have been answered by a mathematician!

Fact:

1 acre $= 4840$ sq yd $= 4840 \times (3)^2$ sq ft $= 43,560$ sq ft.

To the nearest ten thousand, $43,560 \approx 40,000$, and $40,000 = 200 \times 200$, so that a square of side 200 ft has an area of approximately one acre. Since 1 million (10^6) acres is 1000×1000, or $(10^3)^2$, it follows that a square of side 200×1000 ft, or 200,000 ft, has an area of about 1 million acres. Since, to the nearest thousand, there are 5000 ft in a mile (5280 to be exact), this means that our one million acres would be contained in a square with side length about

$$\frac{200,000}{5000} = \frac{2 \times 10^5}{5 \times 10^3} = 40 \text{ miles,}$$

so the area is $\approx (40)^2 = 1600$ square miles. That's quite a large square, but perhaps not as large as one would think, given the headline. The exact answer, by the way, for the length of a side (correct to 2 decimal places) is

$$\frac{(4.356 \times 10^4 \times 10^6)^{1/2}}{5.28 \times 10^3} \text{ miles} \simeq 39.53 \text{ miles!} \text{ The area is } \simeq 1563 \text{ sq miles.}$$

(Furthermore, one acre is contained in a square with side length about 208.7 ft.)

Q.19: Twenty-five billion hamburgers—how many have *you* eaten?

When I first came to live in the United States, I was struck by the huge number of hamburgers sold, as quoted outside one ubiquitous fast-food establishment. I cannot now recall whether that was "Over 100 billion served," but in 1978 it *was* claimed that the "25 billionth" hamburger had been sold. The tradition of keeping track of the number of hamburgers sold has been discontinued, but this leads me to this question: Is 25 billion hamburgers (or meals in general) a reasonable claim for the chain to make?

Since that claim was made three decades ago, let's work with a lower and very convenient value for the U.S. population: 250 million (even though the population in 1978 was closer to 225 million). Then the average number of meals per capita was about 25 billion divided by 250 million, i.e.,

$$2.5 \times 10^{10} \div 2.5 \times 10^8 = 100 \text{ meals.}$$

This seems quite plausible; *on average* a person may have eaten about a hundred meals at "McWhirter's" during the time the chain had been around. This leads to a related question:

Q.20: How many head of cattle would be required to satisfy the (1978) daily demand for meat in the United States?

Again, taking 250 million as the population about three decades ago, let's try and get an upper bound on the estimate by assuming that on average, everyone ate $\frac{1}{4}$ lb of beef each day. Since one "U.S. short ton" is 2000 lb, we can divide the figure in pounds by 2000 to get an estimate in tons. This corresponds to a weight of

$$2.5 \times 10^8 \times \tfrac{1}{4} \div 2 \times 10^3 \approx 30,000 \text{ tons}$$

Now: how many head of cattle does this represent? No cow, surely, could provide a ton of meat, but maybe, say one-third of a ton. Dividing 30,000 by one-third we get an estimate of about 90,000 head of cattle per day at that time! This represents an upper bound because meat eaters do not consume only beef, and not everyone is a carnivore.

Q.21:   Why could King Kong never exist?

He does! I saw him on TV. Perhaps, then, I should give you more information. Let's call him K^2. It seems to me that the producers of that movie may have just taken the proportions of a 5-ft male gorilla and scaled him up in size (maintaining the same proportions) to a creature about 30 ft in height. Now we can use this to think about weight, force, area, and strength, and *then* suggest why K^2 is merely a figment of Hollywood's collective imagination.

We are utilizing ideas of *geometric similarity* here, and for geometrically similar objects made of the same material, their weight is proportional to their mass, their mass is proportional to their volume, and their volume is proportional in turn to the cube of their linear size (think of the volume of a cube, for example; we can think of King Kong as being composed of many tiny little cubes). The strength of an object, in particular K^2, is proportional to the cross-sectional area of his bones required to support his weight, and this in turn is proportional to his area; and since we are dealing with geometrically similar objects his area is proportional to the square of his linear size (think of the area of squares, for example). Note that the word *size* is being used here to indicate a particular dimension, such as height or width (usually the former). When K^2 stands upright, his spine and leg bones have to support his weight. To summarize the argument so far, we know the following.

Weight is proportional to to the *cube* of size; *strength* is proportional to the *square* of size. Therefore, if we take our 5-ft gorilla and scale him up in size very gradually, his weight and strength will both increase, but the former will increase *faster* than the latter (just as h^3 increases faster than h^2 when $h > 1$). So eventually, our gorilla will reach a size where *he is too heavy for his own bones to support him*, and he collapses! (That limit is about 8 to 10 ft in height; it varies with species). This is why *animals cannot grow in size indefinitely without some change in shape* (of course, they cannot grow in size indefinitely anyway!). This is why elephants are not large mice—the legs of an elephant are much thicker in proportion to its body than are those of a mouse. These ideas are developed more generally in the book *Why Size Matters* by J. T. Bonner, cited in the references.

Q.22:   Why do small bugs dislike taking showers?

Have you ever swatted at a fly in the bathroom and wondered where it went, only to find out it has ended up in the toilet? Surely I cannot be the only one to experience that. The fly, once wet, may struggle for ages to extricate itself. Let's think about this case.

When an adult human, Balthazar, who weighs 100 kg (about 220 lb) gets out of a bathtub, he is momentarily covered with a thin film of water about $\frac{1}{2}$ mm thick (the film weighs about 1 kg*). We'll refer to him as B from now on. The thickness estimate of $\frac{1}{2}$ mm comes from observations during my own bath times! We might think of the fly, or any bug for that matter, as a midget-sized person, geometrically similar to the first, but about 1/200th B's size, trying to climb out of the commode with the same thickness of water film covering it. How easy will it be for the fly to climb out? We need to do some sums here, and, by the way, even though big B and the fly are not geometrically similar, that useful simplification will not affect the accuracy of the answer we seek. For B, the ratio of water weight to his weight is 1/100, so 1 kg is a negligible increase in his weight, but for the fly the corresponding ratio is much larger. Since weight is proportional to volume, and hence to the cube of the size, the fly's weight is $(1/200)^3 \times 100$ kg. Its surface area is $(1/200)^2$ times smaller than B's, so the ratio of the weight of water on the fly to its weight is $[(1/200)^2 \times 1$ kg $]/[(1/200)^3 \times 100$ kg$] = 200/100 = 2$. Therefore, the fly will have considerable difficulty getting out of its "bathtub"!

* How do we know that? To estimate these quantities crudely but quickly, consider Balthazar to be a cylinder of radius r and height h: if $r \approx 15$ cm and $h \approx 2$ m, then its surface area $S = 2\pi rh \approx 6 \times 0.15 \times 2 \approx 2$ m^2. Since 1 mm is 10^{-3} m, the volume of this film of water is about 10^{-3} m^3, and since the density of water is 10^3 kg/m^3, its mass is about 1 kg.

Q.23:   How fast is that raindrop falling?

If an object falls from enough height, it will reach a constant speed known as its terminal speed. This is the speed at which the downward force (weight) is balanced exactly by the upward drag force, due to air resistance. Whether the object is a skydiver before opening his parachute, your luggage falling from a plane, or a raindrop, we can use elementary dimensional arguments to show that the terminal speed of a falling object is proportional to $m^{1/6}$, m being its mass. We'll stick with a raindrop since we encounter more of those than anything else, I hope! We'll assume that the atmospheric drag on the raindrop is proportional to the product of its surface area and the square of its speed, a dependence that is pretty well established experimentally.*

Since we are only interested in a proportional relationship, we shall ignore multiplicative constants, and relate drag to a typical drop size L and its terminal speed v_t. As stated above, v_t is attained when the drag force on the raindrop is balanced by its weight. Thus, ignoring all constants, dimensional or not, since weight $\propto$ mass $(m) \propto L^3$, we have that

$$\text{drag} \propto \text{surface area} \times (\text{speed})^2 \propto \text{weight} \propto L^3,$$
$$\text{i.e., } L^2 v_t^2 \propto L^3, \text{ or } v_t \propto L^{1/2},$$

but since $L \propto m^{1/3}$, it follows that $v_t \propto m^{1/6}$. This means that, given two raindrops, one four times the size (diameter) of the other and therefore 64 times as massive, the larger one will only have *twice* the terminal speed of the smaller one. That was easy enough, wasn't it? But of course, it only answers the question in a general way, so let's put in some numbers.

The behavior of small particles like aerosols or tiny cloud droplets falling slowly through air or sediment settling down to the bottom of a lake are well-described by *Stokes' law*, provided their speed of descent is small enough that no turbulence is generated in their wake. This will occur if the *Reynolds number* $\text{Re} = v_t r \rho_a / \mu > 1$, where ρ_a is the density of the air and μ is the dynamic viscosity of the air. For a spherical object of radius r the law states that the drag force F encountered by the object is

$$F = 6\pi r v_t \mu. \tag{23.1}$$

Therefore, if m is the mass of the particle,

$$6\pi r v_t \mu = mg, \tag{23.2}$$

and so

$$v_t = \frac{2\rho_p g r^2}{9\mu}, \tag{23.3}$$

where ρ_p is the density of the particle. Let us first apply this result to calculate the terminal speed of a typical *cloud droplet* of diameter 20 μm (or 2×10^{-5} m) in still air at 5°C at an altitude of 1 km; in the SI system of units $g \approx 9.8$ m/s^2, $\rho_p \approx 10^3$ kg/m^3, and $\mu \approx 1.8 \times 10^{-5}$Ns/m^2. Therefore,

$$v_t = \frac{2 \times 10^3 \times 9.8 \times (10^{-5})^2}{9 \times 1.8 \times 10^{-5}} \approx 1.2 \times 10^{-2} \text{ m/s},$$

or about one cm/s. The Reynolds' number for this droplet is

$$\text{Re} \approx \frac{1.2 \times 10^{-2} \times 10^{-5} \times 1.1}{1.8 \times 10^{-5}} \approx 7 \times 10^{-3},$$

(where the density of the air $\rho_a \approx 1.1$ kg/m^3), so the use of Stokes' law is certainly validated. Note that since $v_t \propto r^2$, a droplet 100 times smaller (of radius 0.1 μm) falls at about 10^{-6} m/s. Conversely, a droplet ten times *larger* (of radius 0.1 mm) falls at about 1 m/s. The Reynolds' number for such a droplet is about ten, so Stokes' law is not valid here, and more sophisticated calculations are necessary. According to an article by A. F. Spilhaus (see references), a raindrop about 2 mm (0.08 in.) in diameter falls at 6.5 m/s (14.5 mph) while a large raindrop measuring 5 mm (0.2 in.) across—the size of a small housefly—falls at around 9 m/s (20 mph). Small drizzle drops fall more slowly, at about 2 m/s (4.5 mph).

---

* This can also be established using "dimensional analysis"; drag, being force (like weight), is proportional to (length)^4/(time)^2, and dimensionally this is realized by the product of area and (speed)^2.

Q.24:   Why can haystacks explode if they're too big? [color plate]

I grew up as the son of a farm worker in England. My father worked on dairy farms, milking cows and making sure that the milk was ready to leave the farm for the next stage in its journey to the consumer. As a young boy, I used to play in the fields near my house, but on one unfortunate occasion, a friend and I set fire to a large barn full of hay (it was entirely his fault, of course). My father was very lucky that he did not lose his job as a result, and I was very lucky that I did not lose my ability to sit down. In the fields surrounding the milking parlor there were many stand-alone piles of hay: *haystacks*.

If they are made too large, they can combust spontaneously, without any help from me. [And you thought spontaneous combustion was just an urban legend!]

But how might such a thing occur? It seems that when the internal temperature of hay rises above 130°F (55°C), a bacterial fermentation reaction begins to produce flammable gas that can ignite if the temperature is sufficiently high, especially if the hay is moist enough. Indeed, according to one website,

> Hay fires generally occur within six weeks of baling. Heating occurs in all hay above 15 percent moisture, but generally it peaks at 125 to 130°F, within three to seven days . . . 150°F (65°C) is the beginning of the danger zone. After this point, check temperature daily . . . [A temperature of] 160°F (70°C) is dangerous. Measure temperature every four hours and inspect the stack. At 175°F (80°C), call the fire department . . . At 185°F (85°C) hot spots and pockets may be expected. Flames will likely develop when heating hay comes in contact with the air. 212°F (100°C) is critical. Temperature rises rapidly above this point. Hay will almost certainly ignite.

Scary stuff. On the basis of this phenomenon, we can pose the following problem for an idealized haystack: *a farmer decided to save time by storing his hay in one very large haystack instead of several smaller ones. Unfortunately, it caught fire so rapidly it appeared to explode! So, can we find the maximum safe size of haystack, given the following information?*

(i) The haystack is a hemisphere of radius $R > 0$; (ii) hay produces heat at a rate of b calories/hr/m^3; (3) the heat escapes at a rate of a calories/hr/m^2. (*Hint:* Consider the heat lost through the surface of the haystack *minus* the heat produced within it, and ignore heat lost through the base.)

Okay. Let the difference of heat lost − heat produced, suggested in the hint, be denoted by $D(R)$, where

$$D(R) = 2\pi R^2(a - bR/3). \qquad (24.1)$$

Subject to an important caveat below, the "safe size" domain for R is that for which the heat lost exceeds the heat produced, i.e., for which $D(R) > 0$, the equality $D(R) = 0$ defining the maximum radius before the haystack is deemed "pyrotechnically unstable"! At this critical value, $R \equiv R_c = 3a/b$. The graph of $D(R)$ is informative and is easily sketched (see figure 24.1); since

$$D'(R) = 0 \Longrightarrow R \equiv R_m = 2a/b, \ D''(R_m) = 0 \Longrightarrow R \equiv R_i = a/b, \qquad (24.2)$$

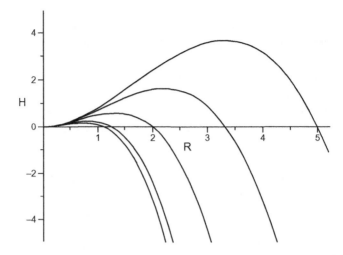

Figure 24.1. The haystack function $H(R)$ for several values of δ

and $D''(R_{\mathrm{m}}) < 0$, $D(R)$ is concave down on (R_{i}, ∞) and attains a unique maximum of $D(R_{\mathrm{m}}) = 8\pi a^3/3b^2$. Thus, $R_{\mathrm{m}} < R_{\mathrm{c}}$ could be termed the safest radius, in that the quantity $D(R)$ is maximized there. Then, from (24.1) above,

$$D(R) = 2\pi R^2(a - bR/3) = 2\pi aR^2(1 - bR/3a) \equiv 2\pi aH(\delta; R), \qquad (24.3)$$

where

$$H(\delta; R) = R^2(1 - \delta R) \qquad (24.4)$$

is the *haystack function*, with the parameter $\delta = b/3a$ being a measure of the heat produced to the heat lost. $H(\delta; R)$ is shown above for several values of δ. For a particular case in which $a = 900$ calories/hr/m^2 and $b = 108$ calories/hr/m^3, $R_{\mathrm{c}} = 3a/b = 25$ m, and $R_{\mathrm{m}} = 2a/b \approx 17$ m.

As already noted, the haystack function is zero for $R_{\mathrm{c}}\ (= \delta^{-1}) = 3a/b$ (ignoring the trivial result $R_{\mathrm{c}} = 0$). In conclusion, then, this simple result means that the difference of heat lost − heat produced per unit time and volume is positive in the interval $(0, R_{\mathrm{c}})$, so more heat is lost than gained; things are stable in this interval. As δ becomes larger, R_{c} becomes smaller, and the range of "acceptable" haystack size is correspondingly reduced. The value of δ for the case mentioned above, 0.04, is not illustrated in the figure; however, these smaller values are probably more realistic. This may be one reason why the old-time haystacks have been replaced by the smaller, rolled-up versions seen frequently in fields and pastures. The calculations for a hemispherical haystack are readily adapted for this geometry (and for a cylindrical roll lying on its side). The conclusions are very similar, apart from geometric factors.

Caveat: The flow of heat from an object to its surroundings, be it a haystack or a cup of coffee, is governed by *Newton's law of cooling*. Somewhat formally, it states that the temperature of an object will change at a rate proportional to the *difference* between its temperature and that of its surroundings. This will be important if the hot object is in an enclosure, because the outside temperature will rise as the object cools, and the temperature rate will be reduced. This will cause an even greater buildup of heat inside the haystack (which has an internal source of heat), and perhaps precipitate combustion even earlier. If, as is more likely, the haystack is in a field, open to the elements, incorporation of Newton's law will not significantly change the overall (admittedly simple) picture presented here. Of course, Newton's law of cooling also applies to objects cooler than their surroundings; both cases are discussed in a more mathematical manner in appendix 3.

In the garden

In this section we start to focus more on real-world problems, having "primed the modeling pump," so to speak. I have learned to notice many things in the garden, while at the same time conveniently missing others, such as the length of the grass . . .

Q.25: **Why can I see the "whole universe" in my garden globe?**
[color plate]

> To see a world in a grain of sand
> And a heaven in a wild flower,
> Hold infinity in the palm of your hand
> And eternity in an hour . . .
> —William Blake, "Auguries of Innocence"

I have a hollow, slightly mirrored glass garden globe, about a foot in diameter. Gazing at it, I can see a reflection of the sky around and above me, and the Earth around and below me—truly, for me, the external visible universe has been compressed (and distorted) to be confined to the surface of my globe! What I see in projection is, of course, a circle, and the images of Earth and sky are compressed more and more toward its circumference.

Suppose the radius of the globe is R, and both I and the objects I see are much further away from the globe than its radius. This is the simplest analytical case, because the light "rays" can be considered parallel. Therefore, any object making an angle θ with the line joining my eye to the center of the globe will appear to be a distance $r = R \sin(\theta/2)$ from the center of the globe (see figure 25.1). Because of the obvious symmetry, we need consider only one-half of the globe. As θ increases from 0 to 180°, r increases from 0 to R, and so, indeed, almost the complete vista of Earth and sky surrounding me is

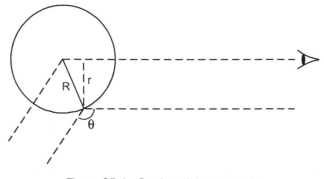

Figure 25.1. Garden globe geometry

imaged on my little globe! The only exceptions are (i) the part behind me, blocked by me, and (ii) the part on the other side of the globe, blocked by my image.

Q.26: How long is that bee going to collect nectar?

Suppose we notice a bee going about garden duties as part of her job description: *collecting nectar*. What might determine how long she stays at a particular flower? This is an interesting exercise in elementary calculus. After the bee finds a flower she sucks it up at a slower and slower rate as the nectar is depleted. Again using anthropomorphic language, let us suggest that she "wants" to optimize the combination of travel time plus feeding time, and ask the question: *when should she give up and move on to the next flower?*

This depends, of course, on how far away the next flower is, but we can make some general observations nonetheless. If she stays too long in one place she will visit fewer flowers, but if she stays only briefly she will spend most of her time flying. Let us further simplify matters by assuming that the flowers are uniformly spaced and hence that her travel time τ between flowers is constant (alternatively we could treat τ as an *average* travel time between differently spaced flowers). We also assume that the bee needs to bring back as much nectar as possible before she "clocks out" for the day. To do this she should maximize the rate of nectar collection per visit (including travel time between flowers). Let $F(t)$ be the amount of nectar collected in time t at a given flower (F is assumed for our purposes to be a differentiable function). In view of the gradually depleting supply, we envisage that this function will be concave down with $F(0) = 0$. We also define $R(t)$ as the rate at which nectar is collected, in the following manner (R is also assumed to be differentiable)

$$R(t) = \frac{\text{food/visit}}{\text{time/visit}} = \frac{F(t)}{t + \tau},$$

where "visit" includes the time to reach the flower (from the previous flower) and the time spent at the flower. It is reasonable to suppose that R should be maximized, which means that we require $R'(t_*) = 0$ and $R''(t_*) < 0$ for some time $t_* > 0$. The first condition implies, after differentiating $R(t)$, that

$$F'(t_*) = R(t_*). \tag{26.1}$$

Figure 26.1 illustrates the meaning of t_* in this context. The second condition will hold for many reasonable forms of F (e.g., a limited growth function like $F = 1 - \exp(-\lambda t)$, where $\lambda > 0$). Indeed, since another straightforward calculation gives

$$R''(t_*) = \frac{F''(t_*)}{t_* + \tau},$$

it follows from the second condition that a local maximum of R occurs at $t = t_*$ provided $F(t_*)$ is *concave down*. But what does the first condition mean to us and imply for the bee? Since R as defined is an average rate of nectar collection over the visit to one flower, it means that when the instantaneous rate of nectar collection equals this average rate, she should move on to pastures new! This type of result is sometimes referred to as the *marginal value theorem*: it has the pithy interpretation for the bee—"Leave when you can do better elsewhere"! These ideas are easily appreciated when illustrated graphically (again, see figure 26.1); note that when τ is smaller, so, too, in general, is the optimal feeding time t_*. For a given value of τ there is a unique value of t_* provided that $F(t)$ is concave downward. Note that in figure 26.1 the slope of the dotted line for arbitrary values of t is $R(t)$, and this is a maximum when it is tangent to the graph of $F(t)$, which occurs when equation (26.1) is satisfied.

In conclusion, we note some interesting facts regarding nectar collection. According to a "Nova" website, it seems that to make one pound of honey, workers in a hive fly 55,000 miles and tap two million flowers; furthermore, in a single collecting trip, a worker will visit between 50 and 100 flowers. She will return to the hive carrying over half her weight in pollen and nectar.

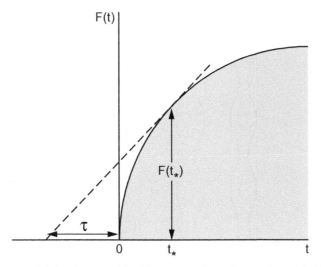

Figure 26.1. The graphical interpretation of equation (26.1)

Theoretically, according to the website, the energy in one ounce of honey would provide our bee with enough energy to fly around the world! At an average speed of only 15 mph, this would take her about 70 days without pit stops! But since a bee weighs only about 0.1 g, and an ounce is 28.25 g, those pit stops would be necessary!

Q.27: Why are those drops on the spider's web so evenly spaced?

Have you ever noticed beads of sticky substance regularly placed along the threads of a spider's web? That substance is the "glue" that spiders use to keep the unfortunate insects that get caught in their webs from seeking pastures new. This "pearling" phenomenon arises, as Philip Ball has noted in *The Self-Made Tapestry*, not because the spider has painstakingly designed it so; rather it is a consequence of the instability of cylindrical columns of liquid to undulations along its surface. The basic idea behind all this is that given any small "waviness" on the surface, the surface tension of the liquid acts to accentuate this curvature, and the result is that each undulation is pulled into a "blob" isolated from its companions. They are strung out like pearls and the unwitting insects are caught, to be captured by the sticky blobs and eaten at a later date. It is interesting to note that the spider only lays down the sticky substance on the spiral threads; the radial ones, constructed first, are not

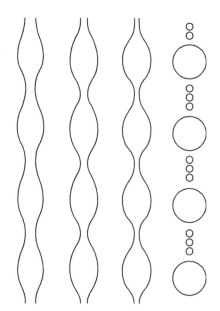

Figure 27.1(a). The "pearling" phenomenon

sticky, so that the spider can move rapidly from the perimeter to the center when lunch arrives.

This instability was studied by Lord Rayleigh at the end of the nineteenth century, and for that reason it is sometimes referred to as the *Rayleigh instability*. It should not be confused with another phenomenon, the *Rayleigh-Taylor instability*, which is associated with adjacent fluids of different densities in a gravitational field. An interesting feature of many fluid-dynamical (and other) instabilities is that while they may occur (as here) for all wavelengths of undulation (or a least a continuous subset), there is usually a particular wavelength λ (or equivalently, a *wavenumber* $k = 2\pi/\lambda$) that is the *most* unstable. This means that wave-like perturbations with this wavelength depart from the cylindrical form (in this instance) most rapidly in time. It is this critical wavelength λ_c that determines the size and separation of the droplets along the spider thread. It is essentially this same instability that is responsible for the breakup into droplets of a thin, nonturbulent stream of water issuing from a tap (see figure 27.1(a)), although the presence of gravity does tend to accelerate the instability by "tearing" the droplets away from the stream. Melting fuse wire (as in the old-time fuses) is subject to the same effects. Ball explains clearly how such pattern-forming processes may be initiated by abruptly occurring instabilities: "Generally an instability sets in when some critical parameter is surpassed . . . Two common aspects of pattern-

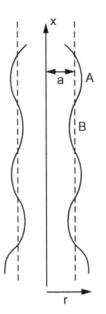

Figure 27.1(b). Notation for equation (27.1)

forming instabilities are that they involve symmetry-breaking . . . and that they have a characteristic *wavelength*, so that the features of the pattern have a specific size." In the present case, the symmetry that is broken is the cylindrical symmetry of the liquid about the axis of the thread.

To discuss this instability in mathematical terms, consider a cylinder of liquid (or soap film) of radius a, and ignore gravitational forces; the fact that the figure is drawn vertically is of no significance. Let the axis of symmetry be the x-axis, and denote the radial direction by r. Now we allow the column to be perturbed by a small periodic or undulatory disturbance of wavelength λ (wavenumber k), so that its radius r is deformed as a function of axial position x in the following manner:

$$r = a + b\cos\left(\frac{2\pi x}{\lambda}\right) = a + b\cos kx, \quad b \ll a. \qquad (27.1)$$

The strong inequality $b \ll a$ means that the disturbance is really small; it changes the ambient radius a by only a small relative amount (this essentially reduces our problem to a linear one as we shall see shortly). The curvature at any point is associated with a pressure, compensating for differences in surface tension. Referring to the diagram of the now-corrugated cylinder (figure 27.1(b)), note that this system is stable (i.e., the uniform cylindrical shape is *restored*) if the pressure at points like A is greater than the pressure at points like B. This is because the pressure will tend to be equalized, and this will induce the opposite curvature at these points, restoring the column towards its original nonperturbed shape.

For any given point on the liquid surface, the maximum possible radial extension under this perturbation is $R_1 = a + b$; in the perpendicular direction (parallel to the axis of symmetry) the radius of curvature (as any calculus book will demonstrate) is

$$R_2 = \left[1 + \left(\frac{dr}{dx}\right)^2\right]^{3/2} \left(\frac{d^2r}{dx^2}\right)^{-1}. \tag{27.2}$$

For sufficiently small initial deformations we may assume that $|dr/dx| \ll 1$, so

$$R_2 \approx \left(\frac{d^2r}{dx^2}\right)^{-1}. \tag{27.3}$$

From the form of the perturbation, it follows that

$$\frac{d^2r}{dx^2} = -k^2 b \cos kx \leq k^2 b, \tag{27.4}$$

so at the point of greatest extension (e.g., A), the *maximum* pressure above the external (atmospheric) pressure is given by the Young-Laplace equation (in terms of the coefficient of surface tension σ)

$$\sigma\left(\frac{1}{R_1} + \frac{1}{R_2}\right) = \sigma\left(\frac{1}{a+b} + k^2 b\right). \tag{27.5}$$

Similarly at points like B, the *minimum* pressure difference above the external is

$$\sigma\left(\frac{1}{a-b} - k^2 b\right). \tag{27.6}$$

Therefore, the pressure difference between points like A and B is

$$\sigma\left(\frac{1}{a+b} - \frac{1}{a-b} + 2k^2 b\right) = \sigma\left(2k^2 b - \frac{2b}{a^2 - b^2}\right) \approx 2b\sigma\left(k^2 - \frac{1}{a^2}\right), \tag{27.7}$$

since by hypothesis $b \ll a$. This pressure difference is positive when $ka > 1$, or $2\pi a = \lambda_c > \lambda$. From our earlier discussion this corresponds to a stable situation; any disturbance with $\lambda < \lambda_c$, tends to damp out. On the other hand, if $\lambda > \lambda_c$, the critical wavelength, then the cylindrical column becomes unstable, causing the sticky thread to fragment into well-defined drops. Thus, instability occurs if the wavelength is too large or the column is too narrow, according to the above criterion. Indeed, Lord Rayleigh showed that the the most unstable wavelength is $\lambda \approx 9a$. It must be pointed out that this analysis is only sufficient to describe the onset of instability; a detailed study of the subsequent instability (for which we have only employed the waving of hands) requires a fully nonlinear analysis. It is interesting to note that this instability can be induced in a narrow stream of water flowing from a tap by placing a vibrating tuning fork close to the stream; if the frequency is low enough, then the corresponding corrugations induced on the surface of the stream will break up into droplets. And this is often what is seen on freshly constructed spider webs! In fact, according to C. Isenberg, the sticky beads are about 10^{-3} cm in radius, though there are two basic drop sizes laid down along the web, one about three times the size of the other. These alternate in a periodic manner, and result from the two wavelengths that characterize the sticky cylindrical tube (associated with each radius of curvature).

Q.28:   What is the Fibonacci sequence?

This question and the next one are designed to set the mathematical scene, as it were, for question 30, which deals with a "fascinatingly prevalent" pattern (to quote H.S.M. Coxeter) present in many plant structures found in your very own garden. To answer this question, however, the Fibonacci sequence is the following sequence of numbers:

$$1, \ 1, \ 2, \ 3, \ 5, \ 8, \ 13, \ 21, \ 34, \ 55, \ 89, \ 144, \ \ldots \qquad (28.1)$$

What do you notice about this sequence of numbers, from the third number onward? Possibly two things, the first being more obvious: each term from the third onward is *the sum of the previous two*; i.e., if f_n represents the nth term in the sequence ($n = 1, 2, 3, \ldots$), then for $n \geq 3$

$$f_n = f_{n-1} + f_{n-2}. \qquad (28.2)$$

That means it's possible to continue developing the sequence forever, unless you have better things to do. Another point to notice is that if you divide each number in the sequence by the next number, beginning with the first, an interesting thing appears to be happening. What is it?

$$1/1 = 1, \; 1/2 = 0.5, \; 2/3 = 0.66666\ldots, \; 3/5 = 0.6, \; 5/8 = 0.625,$$
$$8/13 = 0.61538\ldots, \; 13/21 = 0.61904\ldots,$$

Conversely (and not suprisingly), the reciprocals of these ratios exhibits a correspondingly similar pattern. Thus,

$$1/1 = 1, \; 2/1 = 2, \; 3/2 = 1.5, \; 5/3 = 1.66666\ldots,$$
$$8/5 = 1.6, \; 13/8 = 1,635, \; 21/13 = 1.615\,38\ldots,$$

   The first of these ratios *appears* to be converging to a number just a tad larger than 0.6, and the second to a number greater by one. That, of course, is not a very precise mathematical statement, so we'll endeavor to define these numbers properly in what follows. The expression (28.2) is in fact a linear second-order *difference equation*, second-order because of the difference in the subscripts, in this case, two. It would be very nice to be able to find, say, the 95th Fibonacci number, f_{95}, without having to develop the whole sequence of preceding entries, wouldn't it? Fortunately, help is at hand.

   But let's do a little more mathematics. If we were given a simpler first-order difference equation, such as $f_n = a f_{n-1}$, with $n \geq 1$ say, then clearly $f_2 = a f_1 = a^2 f_0$, $f_3 = a f_2 = a^3 f_0$, etc., so that, in general, $f_n = a^n f_0$. On the basis of this simple concept, a standard method for solving equations such as (28.2) is to substitute $f_n \propto \lambda^n$ into that equation, yielding the quadratic equation

$$\lambda^2 - \lambda - 1 = 0,$$

with roots

$$\lambda_1 = \frac{1 - \sqrt{5}}{2} \equiv \beta \quad \text{and} \quad \lambda_2 = \frac{1 + \sqrt{5}}{2} \equiv \tau. \tag{28.3}$$

Note that

$$-1 < \lambda_1 < 0 \quad \text{and} \quad \lambda_2 > 1,$$

so powers of λ_2 will increase in size, and powers of λ_1 will decrease (though alternate in sign). Note also that

$$\lambda_1 = 1 - \lambda_2 = -\tau^{-1}.$$

The general solution to equation (28.2) is a linear combination of the two particular solutions already found, i.e.,

$$f_n = A_1 \beta^n + A_2 \tau^n \qquad (28.4)$$

where A_1 and A_2 are arbitrary constants. They can be fixed for the Fibonacci sequence

$$(0), \ 1, \ 1, \ 2, \ 3, \ 5, \ 8, \ 13, \ 21, \ 34, \ 55, \ 89, \ \ldots$$

by choosing any two terms as "initial conditions." Note that a term $f_0 = 0$ has been placed at the beginning of the sequence; while it is certainly not necessary to do this, it still permits the difference equation to be satisfied (now for $n \geq 0$) and it has the advantage of greatly simplifying the ensuing algebra. Thus, since $f_0 = 0$ and $f_1 = 1$ the following two equations must be consequences of equation (28.4):

$$A_1 + A_2 = 0 \quad \text{and} \quad \beta A_1 + \tau A_2 = 1. \qquad (28.5)$$

Solving these gives the coefficients as

$$A_1 = -\frac{1}{\sqrt{5}} = -A_2,$$

and so, finally

$$f_n = \frac{1}{\sqrt{5}} (\tau^n - \beta^n). \qquad (28.6)$$

This is called *Binet's formula*. For large values of n this is almost as cumbersome to calculate by hand as using the difference equation (28.2), so a simple approximation to f_n would be very welcome. In fact, this is readily available in view of the fact that $\beta^n \to 0$ as $n \to \infty$, so

$$f_n \approx \frac{\tau^n}{\sqrt{5}} \qquad (28.7)$$

for large enough values of n. An obvious question is—how large is that? Well, judge for yourself. The 14th Fibonacci number, f_{14}, is 377, and $\tau^{14}/\sqrt{5} \approx 377.0006$, which is not bad at all, especially considering we *know* that all the f_n are integers!

Furthermore, from (28.6) it follows that

$$\lim_{n\to\infty} \frac{f_n}{f_{n-1}} = \tau = \frac{1+\sqrt{5}}{2} = 1.61803398\ldots \tag{28.8}$$

This number is called the golden ratio (also the golden number, golden mean, or divine proportion). There are various geometric representations of this fascinating number, some associated with the regular pentagon, but the simplest is found in connection with the line segment below. The Greeks referred to the following as dividing a line in "extreme and mean ratio." The segment AB is divided at C so that the ratio of the larger part (AC) to the smaller part (CB) is the same as the ratio of the whole (AB) to the larger segment (AC). If $AC = \tau$ units and $CB = 1$ unit length, then since

$$\frac{\tau+1}{\tau} = \frac{\tau}{1} \text{ or } \tau^2 - \tau - 1 = 0, \tag{28.9}$$

an equation we have recently encountered. The positive root is again $\tau = (1+\sqrt{5})/2$. Note that $1/\tau = \tau - 1 = 0.61803398\ldots$

$$A\text{————————————————}C\text{————————}B$$

A FIBONACCI DIGRESSION

The 95th Fibonacci number, by the way, is $f_{95} = 31, 940, 434, 634, 990, 099, 905$. That's nearly thirty-two quintillion, or 3.2×10^{19}. Compare that with the estimated insect population of our planet (about one quintillion, or 10^{18}) or the number of different permutations of Rubik's cube (4.3×10^{19}). These are, to say the least, rather large numbers. It would take a long time to count this high! But just for (i) a change of pace, and (ii) fun, I'll *start* the interested reader on estimating just how long . . .

Counting 1, 2, 3, 4, 5, . . . is easier than 946, 383; 946, 384; 946, 385; . . . These longer numbers may take at least 2 seconds to say, so let's take an average of 2 seconds per number, and not allow ourselves any time to eat, sleep, or go to the bathroom (just for the sake of the problem, you understand). For the first million (10^6) of these integers, that's about 2 million seconds. . . so dividing by the number of seconds in a day, $24 \times 3600 = 86, 400$, we have about *23 days*. In counting 1 billion (10^9) numbers, we cannot just multiply 23 days by 1000, because it takes even longer to say numbers like 792, 468, 247. Try it! Those bigger ones take about 3 seconds or even more! And since in

1 billion there are 999 million numbers exceeding 1 million, it follows that there are a *lot* more big numbers than small ones, e.g.,

$$1,000,000,000 - 1,000,000 = 999,000,000, \text{ or } 10^9 - 10^6 = 9.99 \times 10^8.$$

Oh thank heavens for scientific notation! So, at an average of 3 seconds per number, we have 3 billion seconds, or

$$\approx \frac{3 \times 10^9}{8.6 \times 10^4} \text{ days} \approx 35,000 \text{ days},$$

(give or take), or approximately 100 years! Again, no breaks allowed, not even death (which is just nature's way of telling you to slow down). The same problem arises for counting 1 trillion (10^{12}) numbers. Try saying 792,468, 247,139 out loud. It takes me about 5 or 6 seconds, and I speak rather fast. Therefore, since 6 trillion seconds is 2000 times longer than 3 billion seconds, we find this would take about 200,000 years, haha! By now the reader is either asleep, or ready to try these arguments on counting up to a quadrillion (10^{15}) and then a quintillion, which as we have seen, is 10^{18}. Do the sums, but please don't try to verify the estimates by counting in front of your friends.

And one more thing, in case you'd forgotten the original topic; in addition to the golden ratio, there is also a *golden angle* . . .

Q.29: So what is the "golden angle"?

In a circle, the *golden angle* α is subtended at the center by an arc of length "1", that is, the line segment *ACB*. The remaining arc of the circle has length τ. Thus, in figure (29.1) the arc *BA* subtends an angle of, in radian measure, $2\pi - \alpha$, which, of course, exceeds α. Thus, corresponding to equation (28.9) we have that

$$\frac{2\pi - \alpha}{\alpha} = \frac{2\pi}{2\pi - \alpha}, \text{ or } \alpha^2 - 6\pi\alpha + 4\pi^2 = 0, \tag{29.1}$$

yielding a quadratic equation in α with the smallest root being $\alpha = (3 - \sqrt{5})\pi$ radians, or $\approx 137.507°$ (see figure 29.1). Note also that in degree measure, $2\pi - \alpha = 360° - 137.507° \ldots = 222.492° \ldots$ The golden angle, along with the golden ratio, has many applications in botany, one of which, phyllotaxis, is discussed in question 30 . . .

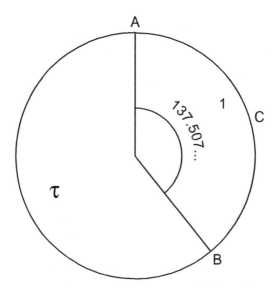

Figure 29.1. Geometry illustrating the golden angle

Q.30:   Why are the angles between leaves "just so"?

Phyllotaxis is the distribution or arrangement of leaves on a stem and the mechanisms that govern it. The term is used by botanists and mathematicians to describe the repetitive arrangement of more than just leaves; petals, seeds, florets, and branches (sometimes) also qualify. These arrangements are closely related to the well-known and previously mentioned Fibonacci numbers 1, 1, 2, 3, 5, 8, 13, 21, 34, 55, 89, . . . and the related golden number or ratio $\tau = (1 + \sqrt{5})/2 \approx 1.618034$ (sometimes the reciprocal of τ, $\tau^{-1} \approx 0.618034$, is referred to as the golden ratio). Numerical and geometric patterns based on these numbers abound in nature and have been studied for hundreds of years, and for that reason alone the basic features of phyllotaxis can be found in many elementary texts.

Examine some flower petals as examples of this. Lilies have 3, buttercups 5, some delphiniums 8, marigolds 13, asters 21, and daisies have 34, 55, or even 89 petals. And there *are* exceptions: the geometer H.S.M. Coxeter has written in his book *Introduction to Geometry* that "phyllotaxis is really not a universal law but only a fascinatingly prevalent tendency." Plants in general face predicaments that we humans can identify with: how to occupy space, receive sunlight, and interact with the environment in an optimal fashion. As a branch grows upward it generates leaves at regular angular intervals, which branch out from the stem. Obviously, if these angular intervals are *exact*

Question 30. Spiral Fibonacci patterns in a pineapple, pinecone, and cactus.

Question 30. (*continued*)

rational multiples of 360°, then the leaves will grow directly above one another in a set of rays (as viewed from above), and would inhibit those below
from sunlight and moisture to some extent (think about leaves sprouting
every 180° ($\frac{1}{2}$ revolution) or 90° ($\frac{1}{4}$ revolution)). In practice, plants seem to
choose rational approximations to the "most irrational" number to optimize
leaf arrangement! We will examine that strange statement a little later on, but
for now we note that, depending on the plant, leaves are generated after
approximately (a very significant word here) $\frac{2}{5}$ of a revolution or circle, (for
oak, cherry, apple, holly, and plum trees), $\frac{1}{2}$ (elm, some grasses, lime, linden),
$\frac{1}{3}$ (beech, hazel), $\frac{3}{8}$ (poplar, rose, pear, willow), and $\frac{5}{13}$ (almond). Other approximations are $\frac{3}{5}$ and $\frac{8}{13}$; these are called phyllotactic ratios, and, as you will
have noticed, their numerators and denominators are Fibonacci numbers
(though not necessarily consecutive ones). An example of a $\frac{3}{8}$ phyllotactic
ratio (along with many other examples) can be found in the book by Garland
(*Fascinating Fibonaccis*); 8 stems are generated in 3 complete turns, not
counting the original stem.

Probably the most striking illustration of phyllotaxis is to be found in the
arrangement of seeds on a large sunflower head. They are distributed in two
families of spirals, perhaps 34 winding clockwise and 55 counterclockwise,
or (55, 89) in the same order, or even (89, 144) in especially large specimens.
Similar patterns occur in daisies also, though, of course, they are more compact
than in sunflowers. Brousseau carried out a detailed examination of such

spirals, particularly in pinecones. Again, there are in general two sets of parallel bract spirals, a steep one from lower right to upper left, and a shallower one from lower left to upper right, perhaps 8 of the former and 5 of the latter, or (3, 5), or (8, 13). According to Brousseau, at least 95% of the spiral numbers in pinecones are from the Fibonacci sequence. Such spirals, as the bracts on pinecones, arise in the petals of artichokes, and also the "scales" of pineapples. In the latter case, three sets of spirals usually occur because the hexagonal-shaped scales have three pairs of opposite sides. The three sets of parallel spirals are usually Fibonacci numbers, e.g., (8, 13, 21), and are oriented respectively at shallow, medium, and steep angles to the axis of symmetry.

The question posed, though, has more to do with the golden angle. The question arises—how do these spirals arise in the first place? To answer this question I shall draw on the very clear explanation given by Ian Stewart in his book *Life's Other Secret*. (He also gives a brief history of the study of phyllotaxis, which we will not go into here). A plant grows upward by generating new cells at the tip of the sunlight-seeking shoot. The potential for spiral formation is already laid down at this stage, as the new cells take up their positions among their neighbors. At the center of the tip is a small circular region of tissue called the *apex*, and around this apex form small lumps called *primordia*. As the apex grows away from the primordia, they "do their own thing" by developing into petals, leaves, seeds, florets, branches, or whatever the case may be. However, the spirals that we see—the *parastichies*—are "merely" by-products of the sequence of events in the plants's growth pattern, a striking optical illusion. The order in which the primordia appear is of fundamental importance here. They trace out a tightly wound spiral—the generative spiral. The further a primordium is from the apex, the earlier it was formed.

Seen from above or from the center of the apex, the angle between successive primordia is always about the same ($\approx$ 137.5°; see figure 30.1). This is called the divergence angle, and is intimately related to the golden ratio τ, as we have already seen and develop below in some detail. For now, however, if we take the ratios of consecutive numbers in the Fibonacci sequence, multiply by 360°, and then subtract that answer from 360° (because we use the internal angle, which is less than 180°), we get a set of approximations to 137.50776 . . .°, which is a more accurate expression of the golden angle (which as we have seen is the mathematical equivalent of an idealized divergence angle). Thus, starting with the pair (3, 5) we can generate the sequence

$$360° \left(1 - \tfrac{3}{5}\right) = 360° \times 2/5 = 144°$$
$$360° \left(1 - \tfrac{5}{8}\right) = 360° \times 3/8 = 135°$$

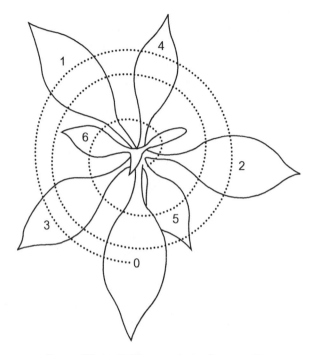

Figure 30.1. Golden angle leaf separation

$$360° \left(1 - \tfrac{8}{13}\right) = 360° \times 5/13 \approx 138.5°$$
$$360° \left(1 - \tfrac{13}{21}\right) = 360° \times 8/21 \approx 137.1°$$
$$360° \left(1 - \tfrac{21}{34}\right) = 360° \times 13/34 \approx 137.6°$$
$$360° \left(1 - \tfrac{34}{55}\right) = 360° \times 21/55 \approx 137.5°$$

Thus, each successive ratio of Fibonacci numbers provides a better approximation to the golden angle. It turns out that if the divergence angle is *less* than 137.5°, gaps appear in the seed head, and only the clockwise family of spirals is seen. Gaps appear again if the divergence angle is *more* than 137.5°, but this time only the counterclockwise family of spirals is noticeable (see figure 30.2). The golden angle thus hits the sunflower on the head, so to speak, for it is the *only* angle at which seeds pack together without gaps, and at this angle both spirals occur simultaneously. As Stewart points out, this most efficient packing makes for a solid and robust seed head. Let us reintroduce the vertical dimension and remind ourselves that every *leaf* also needs its place in the sun, and this is best achieved in a direction not already occupied

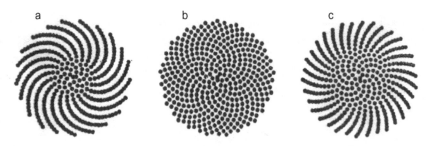

Figure 30.2. Spiral patterns based on the "nearly golden angles" (a) 137.3°; (c) 137.6°. The golden angle 137.5° is used in (b). From Adam (2006)

by a lower one. Thus, as one by one the remaining largest gap gets filled by a socially responsible leaf, things become crowded but in an even and gradual manner, a manner that is possible only with the golden angle as the divergence angle (figure 29.1). As early as 1914, in the first edition of his book, Theodore Cook wrote "the fact that plants express their leaf arrangement in terms of Fibonacci numbers, so frequently that it passes for the normal case, is the proof that they are aiming at the utilisation of the Fibonacci angle which will give minimum superposition and maximum exposure to their assimilating members." While we might disagree with the imputation of motive to the plants, the basic ideas, if not the complete mechanism underlying them, are still quite appropriate today.

We have already noted that any rational multiple of 360° (or 2π radians) will not produce efficient seed or leaf arrangement, merely rays, with gaps between them. For example, if p and q are positive integers, $(360p/q)°$ gives q radial lines. So the multiple of 360° must be *irrational*, i.e., not an exact fraction. But which irrational multiple is best (there are, after all, an infinity of them to choose from)? The quick answer is: one that never settles down to a *rational approximation* for very long, but let us again follow Stewart in his discussion on this. There is a "most irrational" number, and it turns out to be (surprise, surprise) the golden number. The golden number (or its reciprocal, as we have defined it) is the limit of the sequence of ratios of consecutive Fibonacci numbers, i.e., the set of numbers discussed at the beginning of the answer to question 28:

$$\frac{2}{3}, \frac{3}{5}, \frac{5}{8}, \frac{8}{13}, \frac{13}{21}, \frac{21}{34}, \frac{34}{55}, \dots,$$

each new ratio getting closer and closer to $0.6180339\ldots = \tau^{-1}$. These are rational approximations to τ^{-1}. A measure of how irrational a number is can be determined from the errors between the approximations and the number itself. How fast does this difference—the error—shrink to zero? It can be proved that for τ (or its reciprocal) the errors shrink more slowly than for *any* other irrational number. It's the most badly "approximable-by-rationals" number there is! Its "badness" is exceeded only by the awkwardness of the preceding sentence. So the golden number is, indeed, very special, surprising, and strange.

In the neighborhood

Q.31: Can you infer fencepost (or bridge*) "shapes" just by walking past them?

Okay, the question is somewhat unusual, but by *shape* I mean the cross section of each post. I've now left the house behind, and have an unusual *inverse problem* to consider. When walking around the neighborhood or driving in the countryside, you will see various fences on the side of the road, some of which may be made of posts with rectangular cross sections, and others with circular cross sections. Of course, if you are driving at more than 5 mph it's probably not easy to tell one from the other, and why would you want to know anyway? We'll leave one answer to that question until later. Also, at the perimeter of playing fields, stadiums, or sometimes on roofs, one sometimes notices wire netting or mesh, and at a distance the separate wires are indistinguishable from one another, so the mesh looks like a sheet of uniformly gray glass. Looking at it from an increasingly acute angle (relative to the plane perpendicular to the mesh) it may appear to get darker and darker against the background sky. According to Minnaert (1994) this proves (a strong word!) that the vertical wires from which it is made must have a circular cross section (as opposed to being made from small flat bands; see the diagrams). If it were made of the latter, in the form of *very thin* slats, it would remain equally dark at every angle (we could also visualize this question in terms of the slats in a vertical blind). Obviously, the effect of arbitrary rectangular cross sections will affect this last result in general, so let us investigate these statements and try to justify them geometrically. This (and also question 32) is one of those pesky *inverse problems* mentioned in the introduction.

Consider the first situation: small flat bands. Figure 31.1 shows a view from the top in the idealized case of two flat strips of width d and thickness a separated by a gap of width D; this is representative of the whole length of mesh. At a viewing angle θ the corresponding widths of "dark" and "light" for

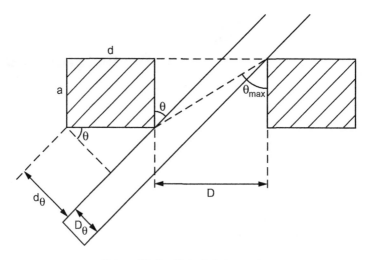

Figure 31.1. Flat slat geometry

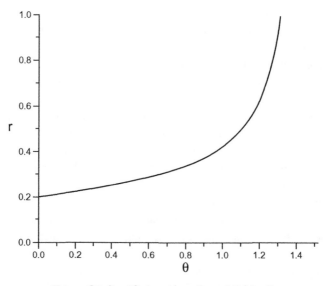

Figure 31.2. $r(\theta)$ for $d/a = 2$ and $D/d = 5$

one band and one gap are respectively $d_\theta + a_\theta = d\cos\theta + a\sin\theta$ and $D_\theta = (D - a\tan\theta)\cos\theta$.

The ratio r of these widths is (see figure 31.2)

$$r(\theta) = \frac{d_\theta + a_\theta}{D_\theta} = \frac{d\cos\theta + a\sin\theta}{D\cos\theta - a\sin\theta} = \frac{d + a\tan\theta}{D - a\tan\theta}, \tag{31.1}$$

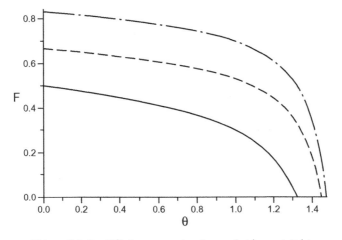

Figure 31.3. $F(\theta)$ for several values of d/a and D/d

provided $\theta < \arctan D/a$. When $\theta = \theta_c = \arctan D/a$ all the light is blocked. Note that $r \to d/D$ as $a \to 0$.

Furthermore, the *fraction of light transmitted* F_θ as a function of θ (for each "gap/post" pair) is best determined by the ratio

$$F_\theta = \frac{D_\theta}{D_\theta + d_\theta + a_\theta} = \frac{D - a\tan\theta}{D + d}, \tag{31.2}$$

which is zero when all the light is blocked, i.e., when $\theta = \arctan D/a$. The graph of this function is shown in figure 31.3 for various values of d/a and D/d:

Strictly, of course, we are looking at the effect of n bands and $n-1$ spaces between them, so we should consider all the contributions, i.e.,

$$\begin{aligned} F_\theta(n) &= \frac{(n-1)D_\theta}{(n-1)D_\theta + n(d_\theta + a_\theta)} \\[2mm] &= \frac{(n-1)(D\cos\theta - a\sin\theta)}{n(D+d)\cos\theta - D\cos\theta + a\sin\theta} \\[2mm] &= \frac{(n-1)(D - a\tan\theta)}{(n-1)(D+d) + d + a\tan\theta}, \end{aligned} \tag{31.3}$$

which approaches F_θ as $n \to \infty$. Note also that as $a \to 0$,

$$F_\theta(n) \to \frac{(n-1)D}{(n-1)(D+d) + d}, \tag{31.4}$$

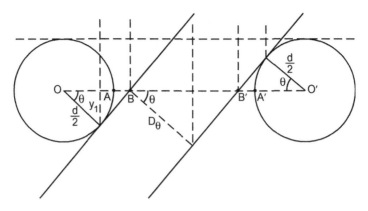

Figure 31.4. Cylindrical post geometry

which is independent of angle θ, justifying the above statement made by Minnaert about very thin bands.

Next we consider the case of cylindrical posts of radius d a uniform distance D apart. For the purposes of analyzing this problem, several distances will be defined here, based on figure 31.4.

Let D be the distance between each post of diameter d. From figure 31.4,

$$D_\theta = BB' \cos \theta = (D - 2AB) \cos \theta, \tag{31.5}$$

but

$$AB = OB - OA = \frac{d/2}{\cos \theta} - \frac{d}{2}, \tag{31.6}$$

so

$$D_\theta = \left(D - \frac{d}{\cos \theta} + d \right) \cos \theta = D \cos \theta + d (\cos \theta - 1). \tag{31.7}$$

When $\theta = \theta_c = \arccos(D + d)$, $D_\theta = 0$ and all the light is blocked between the two posts. The corresponding fraction of light transmitted F_θ (again for each "gap/post" pair) is given by the ratio (see figure 31.5)

$$F_\theta = \frac{D_\theta}{D_\theta + d} = 1 - \frac{d \sec \theta}{(D + d)}. \tag{31.8}$$

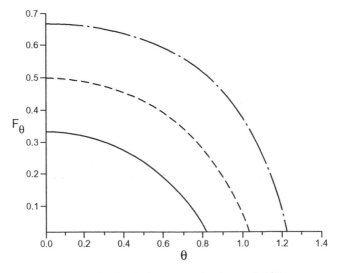

Figure 31.5. F_θ for several values of d/D

This is obviously dependent on θ, unlike the limiting rectangular case discussed above, justifying the above statement made by Minnaert concerning wires with circular cross sections. In a manner similar to that for the rectangular posts, we can show that

$$F_\theta(n) = \frac{(n-1)D_\theta}{(n-1)D_\theta + nd} \to F_\theta \text{ as } n \to \infty. \tag{31.9}$$

So what exactly has been accomplished here? Basically, we have verified some rather broad statements by Minnaert and in the process supplied some further details that were not stated (always a nice feature in modeling). In brief, if wire netting or mesh appears to get darker and darker against the background sky as we move past it, the vertical wires from which it is made must either have a circular cross section, or be made from thick slats. On the other hand, if there is no change in how dark it appears, then they are made of thin slats!

   *Actually, a very good example of this situation arose recently, but with me being stationary, and the "fence" moving. Driving to church on Sunday mornings, Susan and I travel over a bridge that has a central component, made of what seems to be a stiff metal grille in two folding parts. It is hinged at each side, allowing these two pieces to rotate upward and away from each other, thereby permitting ships with tall masts to pass down the waterway. On this particular Sunday morning, we joined a line of traffic halted for this reason, and the bridge was up. I had an excellent view of the scene from where I was

Question 32. A friendly neighborhood pumpkin.

sitting, almost perpendicular to the plane of the large grilles, and soon they were in the process of being lowered (to allow the traffic to move once more). I carefully compared the intensity of the light transmitted with that of the background sky. It was quite amazing; the transmitted component became darker and darker until eventually it was black, at least, from where I was sitting! I must therefore conclude that the metal bridge was composed, in part, at least, of cylindrical metal tubes (or possibly thick bands) oriented perpendicular to the road direction!

Q.32: Can you weigh a pumpkin just by carefully looking at it?

Near Halloween, as I walk around, it seems that every other house has at least one pumpkin near the front door, many of them carved to produce the usual rather grotesque grinning faces. They come in many sizes, and a question that may come to mind is this: *Is it possible to "weigh" a pumpkin by measuring its diameter only?* Let's imagine that we attempt to answer this question in some detail by developing a mathematical model for this problem. Let us further imagine, by way of motivation, that we are dropped in the middle of a pumpkin patch, and wish to find the weight of each one, armed with only a tape measure and our mathematical model, based on the data provided below. And we must not cut any pumpkins open, even during Halloween!

It may be helpful to clarify what we are *not* looking for: this is not intended to be a model describing the pumpkin weight *as a function of time*, but one, in a snapshot of time as it were, that is consistent with the data presented below. If I were to pose this as a set of hints to a formal homework problem it would read something like the following . . .

> State your assumptions clearly, about the shape of the pumpkin, its internal structure (e.g., whether it is hollow or not), the thickness of the shell (if hollow), and so on. It is entirely reasonable to assume that the density of the fleshy part of each pumpkin is (the same) constant. You will probably need to assume a functional relationship of some kind between the thickness of the shell or rim (if hollow) and the "size" (i.e., "radius" or "diameter") of the pumpkin. You may wish to examine a pumpkin of your own and make your own observations. Possible relationships are logarithmic, exponential, power law—it's your choice. If one does not work, try another. You should also try to minimize the number of free parameters in your model. And remember, this is empirical (= "real") data, so expect some data scatter! Also note that, as with most mathematical models, there is no single "correct" answer, but inevitably some answers may be less "correct" than others!"

We'll base our model on the data from a selection of different pumpkins in table 32.1.

ONE APPROACH

Having carved out quite a few pumpkins over the years, I think I've paid my dues, so I'm going to assume that each pumpkin, regardless of size, is a hollow spherical shell with constant shell density ρ. Technically, we need to allow for the density of the air in the hollow region, but it is so small compared with ρ

TABLE 32.1
Pumpkin Data

Radius (cm)	Mass (kg)
14	6
16	10
19	19
21	20
23	26
25	30
30	50

I'll just treat it as zero. The mass of a pumpkin with inner radius R_1 and outer radius R_2 is then

$$M = \tfrac{4}{3}\pi(R_2^3 - R_1^3)\rho, \tag{32.1}$$

and the *weight* is $W = Mg$, where g is the (local) acceleration of gravity, about 9.8 m/s^2. Since weight is proportional to mass, we'll continue to work in terms of M. Let $T = R_2 - R_1$ denote the thickness of the spherical shell; in general, from experience, T is almost certainly a function of the outer radius R_2. Also we'll suppose that the pumpkins are fully grown (and, like adults, come in all sizes), and have varying thicknesses T depending on size R_2. Then, after a tad of algebra used to write $R_1 = R_2 - T$ and to expand R_1^3, equation (32.1) becomes

$$
\begin{aligned}
M &= \frac{4}{3}\pi\rho R_2^3\left[3\left(\frac{T}{R_2}\right) - 3\left(\frac{T}{R_2}\right)^2 + \left(\frac{T}{R_2}\right)^3\right] \\
&= \frac{4}{3}\pi\rho R_2^3[3\delta - 3\delta^2 + \delta^3],
\end{aligned}
\tag{32.2}
$$

where $\delta = T/R_2$. Clearly, $0 < \delta \leq 1$, $\delta = 1$ occurring if the pumpkin is completely solid (a very unlikely case). But we have a little problem regarding T: even if it *were* constant, we don't know its value! Can the data help us? Let's see. It would be nice if δ were small enough that the quadratic and cubic terms could be neglected; from figure 32.1 showing 3δ, $3\delta - 3\delta^2$ and $3\delta - 3\delta^2 + \delta^3$ it appears that for $\delta \lesssim 0.1$ this is reasonable. If we approximate M by retaining both the linear and the quadratic terms, this is reasonable provided $\delta \lesssim 0.3$. As a first approximation, then, let $M \approx 4\pi\rho TR_2^2$ (of course, this could have been written down from the mass of a sphere using differentials). If $T = T(R_2)$, what functional dependence should we pick? Possible dependencies are a power law ($T = aR_2^b$), a logarithmic law ($T = a\ln bR_2$), or an exponential ($T = ae^{bR_2}$); these are simple relationships to consider first, and the simplest of these is probably the power law, so we'll go with that. (It is certainly tempting to suggest that the other forms be "left as exercises for the reader"!)

For convenience we now use 'R' for R_2 in what follows. With the data in the form provided, and for the power law especially, the units for a and ρ are a bit weird, not to mention cumbersome, so let's make everything *dimensionless*. This is a common procedure in science, having as it does the advantage of independence from any particular system of units. To that end, if we take as reference radius (R_0) and mass (M_0) the first entries in the data table, and

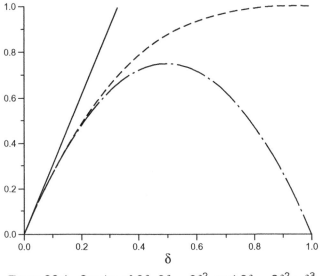

Figure 32.1. Graphs of 3δ, $3\delta - 3\delta^2$, and $3\delta - 3\delta^2 + \delta^3$

define the dimensionless radius $r = R/R_0$ and mass $m = M/M_0$, then we are in a position to rewrite equation (32.2) after making some simplifications. Let's assume that δ is small enough for the linear approximation to be valid. Since in dimensional terms,

$$M = 4\pi\rho TR^2 = 4\pi\rho aR^{b+2} \text{ and, therefore } M_0 = 4\pi\rho aR_0^{b+2}, \qquad (32.3)$$

we now have the simpler looking "model"

$$m = r^{b+2}, \qquad (32.4)$$

for which we need to find an approximate value of b from the data provided. Since m as defined is a measure of relative mass and hence volume, it should come as no surprise that b should be 1 theoretically, and, as noted below, this is borne out quite well by the data. In terms of common logarithms equation (3.4) transforms to

$$\log m = (b+2)\log r, \qquad (32.5)$$

which, of course, is the classic $y = cx$ form of a straight line passing through the origin of coordinates. The graph in figure 32.2, based on table 32.2, shows the points (circles) corresponding to $y = \log m$ and $x = \log r$ and a straight line "eyeballed" through them. Clearly, this would give a statistician nightmares!

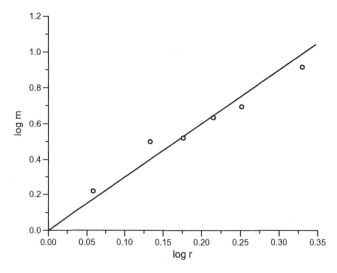

Figure 32.2. Logarithmic data points from table 32.2 and an "eyeballed" linear fit

TABLE 32.2
Normalized Radius and Mass Data

r	$\log r$	m	$\log m$
1	0	1	0
1.143	0.058	1.667	0.222
1.357	0.133	3.167	0.501
1.500	0.176	3.333	0.523
1.643	0.216	4.333	0.637
1.786	0.252	5	0.699
2.143	0.331	8.333	0.921

A proper regression line or least-squares analysis is necessary, but I hope that my eyes are sufficient for the present purposes!

I experimented with the slope of the line and found that $c = b + 2 \approx 3$, so $b \approx 1$ as anticipated. Therefore, the constant a is dimensionless, since both T and R have dimensions of length. We may determine the magnitude of a directly from equation (32.3), i.e.,

$$a = \frac{M_0}{4\pi\rho R_0^3} \tag{32.6}$$

TABLE 32.3
Model-based Thickness T for the Data in Table 32.1

Radius (cm)	T (cm)	Mass (kg)
14	2.4	6
16	2.7	10
19	3.2	19
21	3.6	20
23	3.9	26
25	4.3	30
30	5.1	50

Since M_0 is in kg and R_0 is in cm, both readily measurable units in the pumpkin patch, the density ρ should therefore be expressed in units of kg/cm^3. Like most of us, the pumpkin is made of predominantly water, so we can reasonably replace its density ρ by that of water in these units), i.e., 10^{-3} kg/cm^3. Then with $M_0 = 6$ kg and $R_0 = 14$ cm, we have

$$ a \approx \frac{6}{4\pi} \left(\frac{10}{14}\right)^3 = 0.174, \text{ and so } M = M_0 \left(\frac{R}{R_0}\right)^3 $$

$$ \approx \frac{6R^3}{(14)^3} \approx 2 \times 10^{-3} R^3 \text{ and } T \approx 0.17R. $$

Now we are in a position to add another entry to the original table, based on values of T *predicted* from the model (table 32.3). Here is an exercise for the reader: do a similar analysis for your own set of pumpkins (or ask the neighbors nicely to loan you theirs) and see how accurate the predicted T-values are!

A final point: Obviously the mass of the pumpkin is proportional to the cube of the radius. What we have done here is to find (approximately) from given data, the constant of proportionality in this relationship, and the pumpkin thickness as a specific linear function of the radius. And, in the process, we have engaged in an amusing mathematical modeling project!

Q.33: Can you determine the paths of low-flying ducks?

Most mornings I try to walk a couple of miles at a brisk pace, and being a creature of habit, I tend to walk the same route each time. In fact, it is a distance of two and a quarter miles (as measured by a car odometer) and with no interruptions, it generally takes me about half an hour, corresponding to an average speed of, oh, I'll leave it to you to calculate! (But check your answer in question 93.) Part of the route is along a body of water, and frequently I see ducks, Canada geese, and seagulls on the way, and occasionally heron, cormorants, and even pelicans wheeling their way low over the water. On one occasion two swimming ducks, startled by the speed of my approach, took off and flew away from me obliquely in what appeared to be the same straight line, one slightly behind the other. This is not a particularly unusual occurrence of course, but on this particular morning it set me thinking about how the *angle* subtended at my eye by the ducks varied as a function of time. An odd thing to think about at 7:00 in the morning, perhaps, but no doubt stranger things have happened.

To approach this question, consider figure 33.1. I am the (now stationary) observer, at O, and D_1 and D_2 are the positions, respectively, of the *following* and *leading* ducks at time $t=0$, which I define to be when D_1 (the "following" duck) is at its closest distance to me on its flight path. Their locations at any subsequent time are denoted by D_1' and D_2'. I will make the assumption that they travel at the same speed v, and therefore the distance l between them remains constant. This seems to be a reasonable assumption as the ducks appeared to be mature and of comparable size. Any given difference of speed can easily be accomodated in this little model. Suppose further that the angle $D_1'OD_2'$ subtended at my eye by the ducks is $\theta(t)$ and the angle $D_1'D_2'O$ is $\phi(t)$. We shall use the law of signs in what follows, recalling that in a triangle with vertices A, B, and C, respectively opposite sides of length a, b, and c,

$$\frac{\sin A}{a} = \frac{\sin B}{b} = \frac{\sin C}{c}.$$

Therefore, if the minimum distance between duck D_1 and myself (at $t=0$) is d, then we have from applying the law of sines to the triangle $D_1'OD_2'$ that

$$\frac{\sin \theta}{l} = \frac{\sin \phi}{\sqrt{d^2 + v^2 t^2}}. \tag{33.1}$$

From the larger triangle D_1OD_2' it follows that

$$\sin \phi = \frac{d}{\sqrt{d^2 + (l + vt)^2}},$$

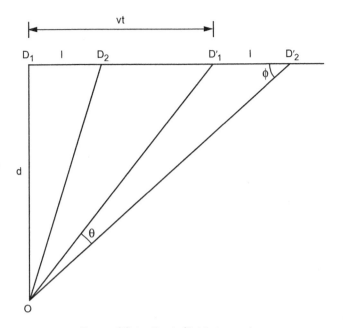

Figure 33.1. Duck flight geometry

so that from equation (33.1)

$$\sin\theta = \frac{l}{\sqrt{d^2 + v^2 t^2}} \frac{d}{\sqrt{d^2 + (l+vt)^2}}. \qquad (33.2)$$

Suppose that $t_0 = d/v$ is the minimum time it would take duck D_1 to reach me along the line $D_1 O$. Then we can introduce the dimensionless "distance" and "time" defined respectively by $\delta = l/d$ and $\tau = t/t_0$. Equation (33.2) may now be expressed in more convenient form as

$$\sin\theta = \frac{\delta}{\sqrt{1+\tau^2}\sqrt{1+(\delta+\tau)^2}}. \qquad (33.3)$$

Note first that, since δ is a constant, the right-hand side of this equation tends to zero like τ^{-2} as t tends to infinity. Of course, I'd be getting very old by then, not to mention hungry, and the ducks, at infinity, would be difficult to see . . . Readers may ask, well, what is the rate of change of θ with time; after all, wasn't that the original question? Not quite, though it's certainly a valid question, and I leave the reader to pursue it. I wanted to see how θ varied

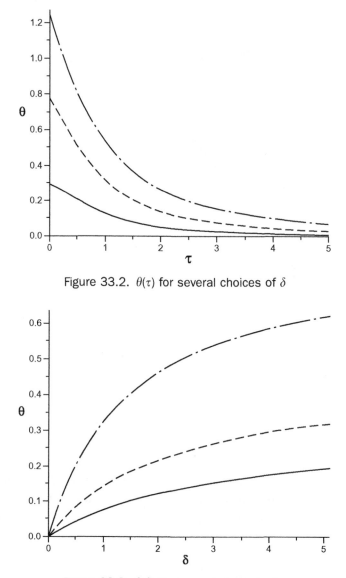

Figure 33.2. $\theta(\tau)$ for several choices of δ

Figure 33.3. $\theta(\delta)$ for several choices of τ

with time, which is not quite the same, and that is readily appreciated by sketching, for several values of the relative separation δ, the angle itself (figure 33.2), i.e.,

$$\theta(\tau) = \arcsin\left(\frac{\delta}{\sqrt{1+\tau^2}\sqrt{1+(\delta+\tau)^2}}\right) \qquad (33.4)$$

The lower curve is drawn for $\delta = 0.3$, the middle curve for $\delta = 1$ and the upper curve corresponds to $\delta = 3$. Clearly, the angle θ decreases monotonically with time in each case, as one would expect, with the angle decreasing most rapidly initially; again there are no surprises. While we are here, it is worth at least looking at the behavior of θ at different times τ, but as a function of δ. In the set of curves in figure 33.3, the lowest is for $\delta = 1$; above that is for $\delta = 2$, and above that for $\delta = 3$. Again, the shape is not surprising; the curves approach the horizontal asymptote $\theta = \pi/2$ as δ increases without bound. This range of δ is quite reasonable because I was quite close to the water's edge, and several duck pairs were involved in this inadvertent experiment!

I wonder what the ducks would say if they knew that the gentleman who startled them was so intrigued by their flight paths that when he got home he started to try and quantify it. I think I know what my family members will say . . .

In the shadows

Q.34: How high is that tree? (An estimate using elliptical light patches)

Using figures 34.1(a) and (b), we may establish a result stated by Minnaert in his delightful book *Light and Colour in the Outdoors* concerning what I call "tree pinhole cameras." The essential idea is that the small spaces between the leaves act as "pinholes," giving rise to the approximately elliptically shaped light patches on the ground. The fact that there is so much shade under a tree, incidentally, is testimony to the effectivness of the foliage; after all, part of its job description is to intercept light and use it in photosynthesis to provide food for the tree. The effects of "leaf shielding" are discussed in question 80.

Suppose a particular pinhole is at height h above level ground, and d is the horizontal distance from a point directly underneath the pinhole to the center of the light patch produced by the pinhole. Assuming this patch is an ellipse of major axis b and minor axis k, the latter subtending a *small* angle θ at the pinhole, the angles α and β in the figure are defined implicitly by

$$\sin \alpha \approx \frac{h}{L} \quad \text{and} \quad \sin \beta \approx \frac{k}{b}. \tag{34.1}$$

Since $\alpha \approx \beta$, we have that $h \approx Lk/b$, and because the sun's angular diameter at the Earth is about $1/108$ radians, we know by similar triangles that $L/k \approx 108$, so

$$h \approx \frac{108k^2}{b}. \tag{34.2}$$

Walking home from work one day along a path lined with crape myrtle trees, I stopped to measure several such light patches, and typical values of b and k were 3 inches and 2 inches, respectively, yielding a value for h of about 12 ft, which looked about right!

Question 34. Leaf shadows, rendered fuzzy by their penumbrae (see plates for questions 36–38 and also question 85 for details on umbral and penumbral shadows).

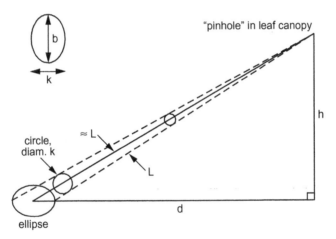

Figure 34.1(a). Geometry for determining tree height from light patches on the ground

Q.35:   Does my shadow accelerate?

As I walk home in the dark, the length of my shadow appears to be increasing faster and faster as I walk at a constant speed v away from a street lamp. *Is this in fact true?* We can use geometric and/or calculus arguments based on

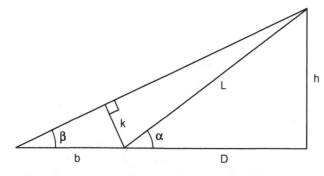

Figure 34.1(b). Light patch detail from figure 34.1(a)

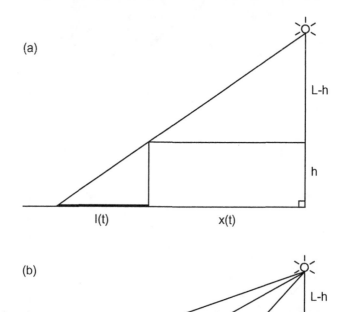

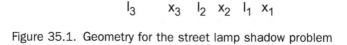

Figure 35.1. Geometry for the street lamp shadow problem

figure 35.1 to prove or disprove this conjecture. The calculus argument is neater in my view, so we'll do that first.

From the diagram, by similar triangles,

$$\frac{l}{h} = \frac{l+x}{L} = \frac{x}{L-h},$$

(35.1)

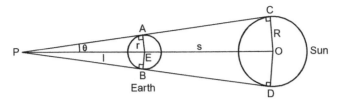

Figure 36.1. Shadow geometry

where l is the length of the shadow, h is my height, x is my distance from the base of the lamp, and L is the height of the lamp. Therefore,

$$l = \frac{hx}{L-h}, \tag{35.2}$$

and using the fact that $v = dx/dt$, the rate of increase of my shadow length is

$$dl/dt = hv/(L-h), \tag{35.3}$$

which is a constant.

The geometric proof of this result is essentially a discrete version of the related rate approach above. If l_n is the length of the shadow after I have walked n units of length x, i.e., a distance nx, then it is readily established that

$$l_n - l_{n-1} = x/(L-h) = l_{n-1} - l_{n-2} = \text{constant}, n \geq 3. \tag{35.4}$$

In fact,

$$l_n = nx/(L-h) = nvt/(L-h), \tag{35.5}$$

so the length of the shadow is proportional to the distance traveled at constant speed, or to the time elapsed. So the apparent acceleration of shadow length must find its explanations in the realm of perception, not physics or mathematics.

Q.36:   How long is the Earth's shadow?

A frequent answer to this question, in class, at least, is "long." Let us try for a little more precision here. From figure 36.1, triangles PCO and PAE are similar, so using the notation in the figure, the length of the shadow $l = rd/R$,

Questions 36–38. The umbral (darkest) and penumbral shadows produced by a pillar partially blocking the light from a street lamp.

where $d = PO$. Strictly, this is measured from the center of the Earth, but as you may suspect, the radius of the Earth is so small compared with that of the sun, this is a very good approximation nonetheless. Since $d = l + s$, it follows that

$$l = \frac{rs}{R - r} \approx \frac{rs}{R}, \tag{36.1}$$

since $R \approx 4.3 \times 10^5$ miles, and $r \approx 4.0 \times 10^3$ miles, so $R \gg r$. The mean Earth-Sun distance s is 9.3×10^7 miles, so putting in the numbers, $l \approx 8.7 \times 10^5$ miles. Almost a million miles! Being rather delighted with this simple calculation, let's now ask the same thing for the Moon, i.e., *How long is the Moon's shadow?* The numerical differences in this calculation obviously arise because now we are dealing with the Moon–Earth–Sun system, so now $r \approx 1.08 \times 10^3$ miles and s is about the same as for the Earth, since the mean Earth–Moon distance (2.4×10^5 miles) is small compared with s. This time, $l \approx 2.34 \times 10^5$ miles, which is just slightly less than the mean Earth–Moon distance. This explains why a total eclipse of the Sun is not guaranteed to be visible anywhere on earth *even if* the three bodies are collinear (which, in general, they are not, even at new and full Moon); under these circumstances only an annular eclipse is guaranteed—the Moon is often just too far away. Fortunately, the Moon's varying distance from the Earth is such that total eclipses *do* occur on occasion! (See question 92.)

Q.37:   And Jupiter's? And Neptune's?

Use the geometric arguments utilized in the preceding question to determine the length l of the (umbral) shadows cast across the solar system by (i) the planet Jupiter, and (ii) the planet Neptune. You may assume that the (approximate) radii of these planets are 4.4×10^4 and 2.5×10^4 miles, respectively, and their respective approximate distances from the Sun are 4.8×10^8 and 2.8×10^9 miles. Express your answer in the same form, i.e., in scientific notation with two significant digit accuracy.

Answer: (i) For Jupiter, $l \approx 4.9 \times 10^7$ miles; (ii) for Neptune, $l \approx 10^8$ miles. Not surprisingly, these are far longer than the shadows in question 36.

Q.38:   How wide is the Moon's shadow?

Of course, this question is quite meaningless unless we specify just *where* the width is to be measured; on the surface of the Earth, of course. We use the information about the (mean) length of the Moon's shadow to calculate the width of the region of totality during a solar eclipse, if the Moon is at its closest distance to the Earth, i.e., $\approx 2.26 \times 10^5$ miles. You may assume Earth is flat over the width of the shadow.

From an obvious variation of figure 36.1 (modified so that r is now the half-width of the umbral shadow, or region of totality, at the surface of the Earth), $r = Rl/(l + s)$, from which the width of shadow is $2r \approx 74$ miles. This is, of course, an upper bound, because the Moon is usually more distant than this. This result is found by neglecting the radius of the Moon compared with its distance from the Earth; a better approximation, without making this assumption, is 72 miles.

In the sky

Q.39: How far away is the horizon (neglecting refraction)?

This is, of course, question 14, "reloaded," and it leads naturally to question 40.

From figure 39.1, we can approximate the distance x from an observer at O, a height h above sea level to the horizon at T. This neglects the effect of atmospheric refraction, of course, but since that increases the effective "distance" by about 9% (see the article by French) this result represents a lower bound. Also ignored is the limiting effect of haze. By Pythagoras' theorem, if the distance $OT = x$, then

$$x = [(R + h)^2 - R^2]^{1/2} \approx \sqrt{2Rh} \tag{39.1}$$

since $h \ll R$. A useful rule of thumb version is easily established if we suppose that h is measured in feet (as is frequently the case) and R is expressed in miles (and more accurately than above, we use $R = 3960$ miles here), then

$$x \approx \sqrt{2Rh} \approx \sqrt{2 \cdot 3960 \cdot \frac{h}{5280}} = \sqrt{1.5h} \text{ miles.} \tag{39.2}$$

Thus, if a tall adult at the seashore looks out to sea, choosing $h = 6$ ft, it follows that $x \approx 3$ miles. On a small hill ($h = 150$ ft), $x \approx 15$ miles. The approximation $h \ll R$ is still valid for an observer at the top of Mt. Everest, so taking $h \approx 29,000$ ft, we find $x \approx 200$ miles. There is quite a view from the top, no doubt (see question 85), although I do not speak from personal experience! For city dwellers, tall buildings suffice to illustrate the point, being closer to home. The roof of the Sears Tower in Chicago is about 1450 ft above ground,

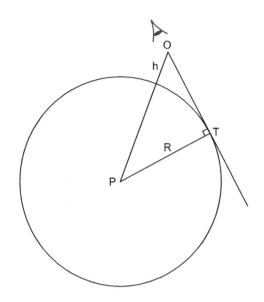

Figure 39.1. Geometry for the horizon distance problem

while that of the Empire State Building in New York City is about 1250 ft above ground. In the ideal circumstances assumed here (perfect viewing conditions), the horizon is close to 47 and 43 miles away as seen from these two buildings, respectively.

But there is more. The selfsame figure can be utilized yet again, this time to determine how far the Moon *falls around* the Earth every second, or, more generally, how far any given planet falls around the Sun in any sufficiently small time interval. We assume the Moon's orbit is circular with a radius $R \approx 2.4 \times 10^5$ miles (this assumption is good enough for the present purposes) and a period of 27.3 days. In fact, these are reasonable mean values. The speed of the Moon in its orbit is, therefore,

$$v = \frac{2\pi \times 2.4 \times 10^5}{27.3 \times 24 \times 60 \times 60} \approx 0.64 \text{ mi/s}. \tag{39.3}$$

So, if the gravitational influence of the Earth were nonexistent at this moment, the Moon would travel tangentially a distance of $x = 0.64$ miles in this first second. Using Pythagoras' theorem as before,

$$h \approx \frac{x^2}{2R} \approx 8 \times 10^{-7} \text{mi} \approx \frac{1}{20} \text{th inch}. \tag{39.4}$$

Question 40. A small cumulus cloud moving across a hill in Banff National
Park, Alberta. A lenticular wave cloud can be seen just above the brow of
the hill.

Q.40:   How far away is that cloud?

Let's get specific, using the approach of the answer to question 39. A person
standing on top of a hill 800 ft high sees a cloud on the (unobscured) horizon.
If, from her vast knowledge of meteorology, she knows the minimum altitude
of the cloud to be 1 mile, what is the minimum possible distance from the observer to the cloud? (See figure 40.1.)

From the (greatly exaggerated) diagram, drawn for an arbitrary value of h,
and by the application of the result of the previous question, $OT \approx \sqrt{1.5\,h}$
miles, where $h = 800$ is in feet, i.e., $OT \approx \sqrt{1200} \approx 35$ miles. This is the distance to the horizon (neglecting the effects of refraction). Obviously, the cloud
cannot be as high as it is and still be nearer to the observer than the horizon

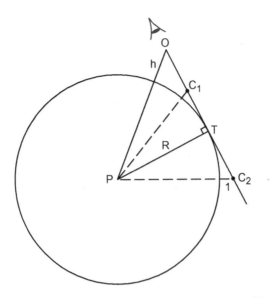

Figure 40.1. Geometry for the cloud distance problem

(at distance OC_1), so it must be in the line of sight beyond the horizon, at a distance of at least

$$OC_2 = OT + TC_2 \approx \sqrt{1200} + \sqrt{(R+1)^2 - R^2} \approx \sqrt{1200} + \sqrt{2R}$$
$$= \sqrt{1200} + \sqrt{7920} \approx 124 \text{ miles}, \qquad (40.1)$$

where R is the radius of the earth. In practice this would not be visible because of the lack of contrast due to atmospheric extinction (see Bohren and Fraser 1986).

Q.41: How well is starlight reflected from a calm body of water?

Sometimes early in the morning, or later in the evening, I notice reflections of the Moon or streetlights from the surface of bodies of water as I walk around them. These reflections are usually oblique, of course; the Moon or street-lamps are not directly overhead as I view their reflections. According to Greek legend, the young man Narcissus was so proud and haughty that he angered the gods, and they condemned him to fall in love with his own reflection in a pool of crystal clear water. He was so enamored of himself that he could not bear to leave the pool of water. Let's do some mathematics

TABLE 41.1
Reflectivity of Light as a Function
of Zenith Angle

Zenith distance (degrees)	% Reflectivity (reflectance)
0	2.0
10	2.0
20	2.0
30	2.1
40	2.4
50	3.5
60	5.9
70	13.2
80	34.7
90	100

with this gentleman; of course, for now we are dealing with the simpler case of direct (or normal) incidence: looking down directly upon the surface of the water.

From a study of Maxwell's equations of electromagnetic theory (applied to the reflection and refraction of waves at a plane interface) it can be shown that the *reflectivity* (or *reflectance*) R of light normally incident on a flat body of water is given by

$$R(n) = \left(\frac{n-1}{n+1}\right)^2, \tag{41.1}$$

where R is the fraction of the incident intensity that is reflected at the air/water interface (and also the water/air interface, going in the other direction); n is the refractive index of water (we may take $n = 4/3$). If light is reflected from below the surface (i.e., from the bottom of the pool), the refractive index is then n^{-1} or $3/4$, but of course R remains unchanged for normal incidence because $R(n) = R(n^{-1})$. (For values of R in other cases, see table 41.1.)

Questions: (i) Consider first the case of *zero reflectivity* at the bottom of the pool (all the light reaching there is absorbed by the black disgusting ooze and who-knows-what-else that is buried there). What fraction of the original light intensity is reflected back to Narcissus? (ii) Next consider the

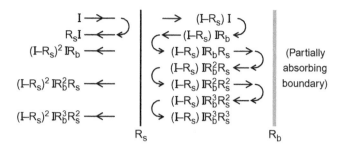

Figure 41.1. Intensity reflection and transmission in a pool of water

bottom of the pool to have a reflectivity of 20% (perhaps there is a piece of glass or other reflecting material lying there). *Now* what fraction of the original light intensity enters his eyes? Is he likely to see his own reflection? Why or why not? You may assume that there are no other extraneous relections to interfere with what he sees. What should the bottom reflectivity be in order that the light reflected from the surface of the water and the light entering his eyes from the bottom are the same intensity?

We shall examine these questions in two stages: approximate and then, more briefly, the exact formulation. We will denote by R_s the reflectivity of the surface of the water, and by R_b the reflectivity of the bottom of the pool. If the intensity of light incident normally (from the face of Narcissus) is I, then the intensity reflected from the surface is $R_s I$ and that transmitted is $(1 - R_s)I$; the proportion of the latter term reflected from the bottom of the pool is $R_b(1 - R_s)I$, and the proportion of *this* latter term transmitted is $R_b (1 - R_s)^2 I$, etc. This procedure can be continued indefinitely, as indicated in figure 41.1.
    Now

$$R_s(4/3) = R_s(3/4) = 1/49 \approx 0.02,$$

so if the bottom has zero reflectivity ($R_b = 0$), then the reflected intensity of Narcissus' face is only about 2%. Regarding the second part of the question, if $R_b = 0.2$, the first return transmission gives $R_b (1 - R_s)^2 I \approx 0.19I$, or about 19% of the of the original intensity; this would completely swamp the 2% reflected intensity discussed in part (i) of this question.
    Let's apply this to starlight reflected in water. As discussed later in question 87, the observed brightness of stars is expressed in terms of their *apparent magnitudes m* on a numerical scale that increases as the brightness decreases:

$$m = 6 - 2.512 \log_{10}\left(\frac{L}{L_0}\right) \tag{41.2}$$

where L is the light flux (luminosity or brightness) of the star (or planet) and L_0 is the brightness of the faintest star visible to the (average human) naked eye. So if light from an overhead star of magnitude m is reflected in a body of water, with a 2% reflected intensity, i.e., $L/L_0 = 0.02$, then the apparent magnitude of the reflected starlight is (see question 87)

$$m_{\text{refl}} = m - 2.512 \log_{10}(0.02) \approx m + 4.3, \qquad (41.3)$$

so a star of the magnitude 1.0 appears in reflection to be of the 5th magnitude. If the planet Mars or Jupiter with magnitude -2, say, were reflected in this manner from water, their reflections would be well in the 2nd magnitude range. The reflectivity of water is considerably higher for stars closer to the horizon; at about $40°$ above the horizon, $R \approx 0.06$; at about $5°$ above the horizon, $R \approx 0.6$, though these values do not take account of the increased absorption of starlight from stars of a given magnitude because of the longer path through the atmosphere the light has to traverse compared with higher sky locations.

Comment 1: Of course, the details of part (i) are just an approximate approach to the problem and are probably sufficient for most purposes, but the complete problem (when $R_b \neq 0$) involves an infinite number of reflections and transmissions at the surface of the pool because of the partially reflecting boundary at the bottom of the pool. Based on the diagram, we can calculate the total proportion of light intensity moving toward Narcissus *above* the water: it is

$$I_L = R_s I + R_b (1 - R_s)^2 I [1 + (R_b R_s) + (R_b R_s)^2 + (R_b R_s)^3 + \cdots].$$

The total proportion of intensity moving away from Narcissus *within the water* (i.e., excluding the initial intensity) is

$$I_R = (1 - R_s) I [1 + (R_b R_s) + (R_b R_s)^2 + (R_b R_s)^3 + \cdots].$$

Both of these right-hand sides are geometric series, with

$$[1 + (R_b R_s) + (R_b R_s)^2 + (R_b R_s)^3 + \cdots] = (1 - R_b R_s)^{-1},$$

reducing to the results for part (i) when $R_b = 0$.

Comment 2: We have not been very adventurous by merely playing with equation (41.1). We could ask where this comes from, and refer it back one step to the comment about Maxwell's equations; the equation describing reflectance for *arbitrary* angles of incidence (*Fresnel's equation*) is in fact

$$R(i, r) = \frac{1}{2} \left[\frac{\tan^2(i - r)}{\tan^2(i + r)} + \frac{\sin^2(i - r)}{\sin^2(i + r)} \right], \tag{41.4}$$

where i is the angle of incidence, r is the angle of refraction, and, of course, by Snell's law of refraction,

$$\sin i = n \sin r. \tag{41.5}$$

When the Sun or a star (or Narcissus' face) is directly overhead, $i = r = 0$, but in the limit of very small angles, $i \approx nr$, so by substituting this in equation (41.2) and taking the formal limit, equation (41.1) is recovered. We can use (41.2) to provide a table of R for various angles of incidence (or, equivalently, zenith angles; measured down from the point directly overhead).

This table (based on a slightly more accurate choice of $n = 1.34$ for water) indicates that the reflectivity is quite insensitive to zenith angle, at least initially, and obviously increases rapidly as the Sun or star approaches the horizon, corresponding to near tangential reflection. Again, this is all for a perfectly smooth surface, something that is rarely likely to occur in practice. Bodies of water large enough to see shallow angle reflections from stars near the horizon are rarely ever really still. This means that in practice the images of those low stars are very difficult to see.

Q.42: How heavy is that cloud?

What about the weight of a cloud? According to Jack Williams, author of *The Weather Book*, meteorologist Margaret LeMone "weighed" a smallish cumulus cloud as it drifted over the Plains east of Boulder, Colorado, helped by the fact that from the Appalachians westward, the United States is divided into squares one mile on a side, usually marked by roads. This particular cloud had a shadow about 0.6 mi (or about 1 km) square, and it was about as high as it was wide. This corresponds to a volume of about $1 \, \text{km}^3$ or $10^9 \, \text{m}^3$. Typically, each cubic meter of such a cloud contains about half a gram of water in liquid form, so this one had a total of about $5 \times 10^8 \, \text{g}$, equivalent to 500 metric tons. For purposes of comparison, a Boeing 747-400 jetliner with a full load of passengers (up to 524) and fuel has a maximum takeoff weight of about 400 metric tons. So even a moderate-sized puffy cloud weighs more than a fully-laden

Jumbo! Of course, there would be a bit more room to move around in an aircraft the size of a cloud! Thus, prorating this result for a small cloud, say 50 m across, gives a weight of about 60 kg; doing the same thing for an Olympic swimming pool, with a volume of about $50 \times 25 \times 2 = 2500 \text{ m}^3$, which weighs 2500 metric tons, about five times the weight of Margaret's cloud, and $(5)^{1/3} \approx 1.7$ times as large linearly, or about a mile across.

Q.43:   Why can we see farther in rain than in fog?

We start off with some simple definitions. Suppose that the diameter of a spherical water droplet in centimeters is denoted by L. Further suppose that 1 g of water (volume: 1 cm^3) is dispersed in the form of a mist or rain shower. This produces N identical droplets, where

$$N = \frac{\text{volume of water}}{\text{volume of a drop}} = \frac{1 \text{ cm}^3}{\pi L^3/6 \text{ cm}^3} \approx \frac{2}{L^3}, \qquad (43.1)$$

Thus, N is a function of the drop size L, and so we find that if L is in centimeters, $N(10^{-1}) \approx 2 \times 10^3$; $N(10^{-2}) \approx 2 \times 10^6$ and $N(10^{-3}) \approx 2 \times 10^9$. Now let V be the *relative volume* of water present in a given unit volume of air (e.g., 1 m^3). The above volume of water yields $V = 10^{-6}$, which is about right for a heavy mist or rain, according to the book by Vergara (see references). Now expressing L in meters, we have

$$N \approx \frac{2V}{L^3}, \qquad (43.2)$$

and assuming there is no overlap of drops in the line of sight, each drop will block off an area

$$A = \pi \left(\frac{L}{2}\right)^2 \approx \frac{3}{4} L^2 \text{m}^{-2}. \qquad (43.3)$$

(If there *is* any overlap, this represents an upper bound to the blocked off area.)
    The *total area* blocked off is

$$N \times A \approx \frac{2V}{L^3} \times \frac{3}{4} L^2 \approx 1.5 \frac{V}{L} \text{m}^{-2}, \qquad (43.4)$$

which implies that, for the same relative volume, *smaller drops block off a larger area*, i.e., are cumulatively less transparent than a cloud of larger drops. Let's put some numerical flesh to this skeleton: imagine looking through a linear stack of s one-meter cubes, with total volume $s\,\mathrm{m}^3$ and containing total water volume Vs. The total area blocked off is, as an upper bound,

$$A_{\text{total}} \approx 1.5\frac{Vs}{L}. \tag{43.5}$$

Now suppose that there is only 10% visibility, i.e., 90% of the viewing area is blocked off. How "long" is the fog bank or rain shower? Well,

$$0.9 = 1.5\frac{Vs}{L}, \tag{43.6}$$

and using our typical value of $V = 10^{-6}$, we find

$$s = 6 \times 10^5 L. \tag{43.7}$$

For large drops $(L \approx 1\,\mathrm{mm} = 10^{-3}\,\mathrm{m})$, $s \approx 600\,\mathrm{m}$; For fine mist $(L \approx 10^{-2}\,\mathrm{mm} = 10^{-5}\,\mathrm{m})$, $s \approx 6\,\mathrm{m}$. That seems to accord with my experience; how about yours?

Q.44: How far away does that "road puddle" mirage appear to be?

Does light travel in straight lines, line segments, or curves? Well, it all depends on whether the *speed c* of light in the medium varies. Generally, it (or the speed of sound, for that matter) is not constant as it propagates in a medium. The *refractive index* of a medium, if constant, can be defined as the ratio of the speed of light in a vacuum to the speed in the medium. Typically for water, as noted in question 41, it is about $4/3$ and for glass about $3/2$, though n is slightly wavelength-dependent (without which we would have only "white-bows" instead of rainbows). If the refractive index n varies in a given manner, $n = n(y)$, say, where y is the altitude above the surface of a flat Earth (for simplicity here), then we can write c in terms of $n(y)$, such that

$$c(y) = \frac{c_0}{n(y)}, \tag{44.1}$$

where c_0 is (for light) the speed of light in a vacuum (about 186,000 mi/s or 300,000 km/s). It is the variation of n (and hence c) that gives rise to mirages.

Mirages are fascinating and, by definition, deceptive to the casual observer. Perhaps no better example of this is afforded by the writings of Robert E.

Peary, who in 1906 (en route, as he hoped, to the North Pole), stood on a summit and saw to the northwest at a distance of about 120 miles (as he believed) "snow-clad summits above the ice horizon." This mysterious yet inviting "land" was eventually named "Crocker Land," and in 1913 an expedition, led by Donald B. MacMillan, set out to find and explore it. As they approached its apparent location, he wrote, "There could be no doubt about it. Great heavens, what a land! Hills, valleys, snow-capped peaks extending through at least 120 degrees of the horizon." After moving some 30 miles toward this fantastic land, they found nothing. Some might call this a cruel hoax played on them by the laws of optics; but it was not an optical illusion, for the image was real enough—it just did not coincide with an object. It was a *mirage*, and that is one convenient way of defining the phenomenon.

The mirage of Crocker Land witnessed by Peary and MacMillan was probably an example of what has come to be called the *Fata Morgana*— possibly the most spectacular member of the entire class of mirages. It is the Italian name for the Fairy Morgan, who, as legend has it, was King Arthur's sister, and she possessed, it seems, the ability to create castles in the air (as some people claim mathematicians are wont to do; but at least we then inhabit them). The Italian connection comes from a description of such "castles" written by a priest, Father Angelucci, in a letter to a colleague concerning his observations on the morning of August 14, 1643, while looking across the Strait of Messina toward the island of Sicily. He saw what appeared to be a dark mountain range in front of which appeared a variety of different images, including columns, arches, towers, and windows. (References and further details of these events may be found in the article by Fraser and Mach.)

We will not attempt to explain the various different types of mirages here, being content to answer the original question, *How far away does that "road puddle" mirage appear to be?* in a straightforward context. We will examine the common "water on the road" *inferior* mirage, so called because it is an image lying below the object producing it (in this case, a portion of the sky). To set the scene, we are walking along the side of the road on a warm day and witness this common sight. What has happened is that the road and the adjacent layer of air above it have been warmed by the Sun, and this air is less dense than at higher levels, where the air is somewhat cooler. Because the weight of a column of air changes only slowly with altitude, the air pressure is approximately constant across these layers. We'll make use of Snell's law of refraction (equation (41.5) above), generalized to the case of a continuously varying medium in the form

$$n(y) \sin \theta(y) = \text{constant} = D, \tag{44.2}$$

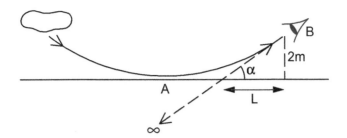

Figure 44.1. Inferior mirage geometry

where θ is the continuously varying angle that the light "ray" makes with the vertical (or y) direction (we are considering a flat Earth here). From equation (44.1) this equation can be rewritten as

$$\sin \theta(y) = Kc(y), \tag{44.3}$$

K being another constant. To interpret these equations physically, suppose that, as in the example we are considering, the density of the air increases (in the vicinity of the road surface) with height y above the road. Then so does the refractive index n, and from (44.2), the angle θ must decrease, so the ray path steepens and is concave upward. Equivalently, (44.3) implies that the speed of light decreases upward. (The reverse would occur if the air density decreased with altitude; the resulting ray paths would then curve downward.)

We'll also need to express the relationship between density and temperature (also known as the ideal gas law) in terms of the refractive index. This formulation is called the *Gladstone-Dale law*, and it states that, in a gas, the variables $(n-1)$, ρ and T^{-1} are all proportional (the symbol $\propto$ is used to mean "is proportional to"), i.e.,

$$(n-1) \propto \rho \propto T^{-1}, \tag{44.4}$$

where ρ is the density of the gas at any point and T is the corresponding *absolute* temperature (remember, the pressure is approximately constant). The quantity $n-1$ is the difference between the refractive index at a point in space and its value in a vacuum. Note that the conditions described in this example will give rise to an inferior image because the temperature *decreases upward*, the density *increases upward*, and so the ray path is concave upward, as shown in figure 44.1. In fact, there is a "mirage theorem" stating that the concavity of the ray at any point on its path is always toward regions of higher density.

To get specific, on a day such as that described here, the air temperature six feet, or about two meters, above the road might be $T_B \approx 30°C$ (86°F) and near the road surface $T_A \approx 50°C$ (122°F). At these temperatures, the refractive index at your eye may take the value $n_B = 1.00026$. We wonder how far away the mirage appears to be on the road . . .

We make use of the fact, mentioned above, that $n \sin \theta = D$, and the Gladstone-Dale law (44.4), but with a caveat. We are now going to approximate the continuous variation in n and T by a two-layer model, in which both quantities have different but constant values, n_A, n_B and T_A, T_B. From the (greatly exaggerated) sketch in figure 44.1, the angular deviation of the ray entering the observer's eye from the horizontal is α, so from Snell's law, equation (44.2),

$$n_B \times \sin (\pi/2 - \alpha) = n_A \times \sin (\pi/2), \tag{44.5}$$

i.e.,

$$\cos \alpha = n_A/n_B \approx 1 - \alpha^2/2, \tag{44.6}$$

(since generally $\alpha \ll 1$), where α is in radians, and we have approximated $\cos \alpha$ by the first two terms in its Taylor series expansion. From this expression and the Gladstone-Dale law in the form

$$(n_A - 1)T_A = (n_B - 1)T_B, \tag{44.7}$$

we find that

$$\alpha^2 \approx 2(1 - n_B^{-1})(1 - T_B/T_A). \tag{44.8}$$

With temperatures expressed in Kelvins, $T_A \approx 323$ K, $T_B \approx 303$ K, and n_B as given, $\alpha \approx 5.7 \times 10^{-3}$ radians, or $\approx 0.32°$. Since $\alpha \approx \tan \alpha = 2/L$ when α is in radians, it follows that the distance to the point where the mirage appears to be on the road is $L \approx 2/\alpha \approx 350$ m (1150 ft). Does that seem about right to you?

In closing this question, we might note that two of the most common mirages are rarely recognized as such: sunrises and sunsets. The apparent distortion of the solar disk is clearly due to differences in refractive indices in the air near the horizon, but when the lower limb (or edge) of the Sun has just broken above the horizon, the Sun's disk is actually still below it! Conversely, when the lower limb is just grazing the horizon, the Sun has already set. And this is independent of the changing colors of the Sun due to differential refraction and absorption of the solar spectrum as it nears the horizon. Indeed, the rather rare phenomenon known as a *green flash* is due in part to these effects. According to his website (see references), atmospheric scientist Andrew

Young writes that "Green flashes are real (not illusory) phenomena seen at sunrise and sunset, when some part of the Sun suddenly changes color (at sunset, from red or orange to green or blue). The word 'flash' refers to the sudden appearance and brief duration of this green color, which usually lasts only a second or two at moderate latitudes." Furthermore, in explaining the phenomena, he states:

> The basic cause of the color is atmospheric dispersion: refraction by air is larger at shorter wavelengths. So, at sunset, the refractive delay of the sunset is usually a second or two longer for blue and violet than for red. In general, then, the red image of the Sun (or of some miraged part of it) sets or disappears first, followed by yellow, green, blue, and violet. So why isn't violet the last color to be seen at sunset? There is another effect at work: atmospheric extinction. Both air molecules and aerosol particles scatter the shortest wavelengths most strongly . . . At the horizon, the path length through the air is very long, and the shortest wavelengths are almost completely removed. Scattering by molecules alone is not quite enough to make the shortest wavelengths invisible; so if the air is very clear, violet is the last color seen. But usually there is enough haze in the air that violet, and even blue, are completely removed, so that green is the last color seen at sunset, or the first at sunrise.

The reader is strongly encouraged to explore Young's website, which provides a comprehensive account of the history and erroneous "explanations" of green (and other) flashes, together with detailed scientific discussion of them along with many other links to related phenomena, such as—no surprise here—mirages.

Q.45: **Why is the sky blue?** [color plate]

In a delightful book by Robert Ehrlich, entitled *The Cosmological Milkshake*, there appears a cartoon showing a child sitting on a bench with her father; the caption to this picture reads: "The budding urban scientist asks, 'Daddy, why is the sky brown?'" Returning to the original question, note that sunlight entering the Earth's atmosphere is scattered by the molecules in the air, which are small compared with wavelengths of light. The electric field of the incident sunlight causes electrons in the molecules to oscillate, and re-radiate the light; this is what is meant here by the word "scattering." The degree of scattering is inversely proportional to the *fourth power of the wavelength* of the light; blue light being of shorter wavelength than red, it is scattered the most. Consequently, we see blue sky except when we look in the direction of the sun at sunrise or sunset, when the long path through which the light passes depletes the blue light, leaving a predominance of the longer wavelength red light. This phenomenon, coherent scattering, is often referred to as *Rayleigh scattering*, named for Lord Rayleigh, who developed the theoretical basis for this type of scattering. Sunset colors are also determined by the amount of dust (or aerosols in general) in the atmosphere; after major volcanic eruptions they can be really spectacular, as are the sunrises, and can occasionally give rise to so-called "blue moons" (but only once in a blue moon).

To see how this result comes about, consider the following simplified argument. It gets a little technical, but stick with it! Because it depends on many variables, we let the intensity I of the radiation scattered from a particle of volume V be described by the expression

$$I = f(V, r, \lambda, n_1, n_2, I_0)$$

where r, λ, n_1, n_2, I_0 are respectively the distance r from the scatterer to the observation point, the wavelength λ of the scattered radiation, the refractive indices of the exterior and interior media (n_1 and n_2, respectively), and the intensity I_0 of the incident radiation. The refractive indices, being ratios of speeds of light in different media, are dimensionless and, to be specific, we seek the following algebraic form:

$$I \propto V^\alpha r^\beta \lambda^\gamma I_0$$

Because volume has dimensions of (length)^3, which is written as $[L]^3$, and both r and λ have dimensions of length also, a particularly simple result follows by equating the dimensions of both sides of this equation, to obtain

$$[L]^{3\alpha}[L]^\beta[L]^\gamma = [L]^0$$

because the dimensions of intensity "cancel" from both sides. Since the amplitude of scattered radiation is proportional to the number of scatterers, which in turn is proportional to the volume of the (composite) particle, and intensity is equal to the square of the amplitude, we have that $\alpha = 2$. Furthermore, $\beta = -2$ since a dipole (e.g., a molecule in which the positive and negative charges are not coincident) radiates energy in all directions (think of the surface of an expanding sphere of light; the energy per unit area decreases as (radius)$^{-2}$). This means that $6 - 2 + \gamma = 0$, or $\gamma = -4$. Thus, in particular,

$$I_{\text{scatt}} \propto \lambda^{-4}$$

which is effectively *Rayleigh's inverse fourth-power law of scattering*. Rayleigh scattering theory applies if the particles are small, i.e., $\lesssim 0.1\lambda$; under these circumstances they can be considered as point dipoles. Much larger than this size, the light scattered from one part of the particle may be out of phase with that from another part; the resulting interference reduces the intensity of the scattered radiation. There is now a greater intensity in the forward direction, because the cumulative effect of the phase differences is smallest for small scattering angles. *Mie scattering theory* takes into account all these features and, using Maxwell's equations of electromagnetic theory, the scattering intensity can be expressed as an infinite series of so-called *partial waves*; Rayleigh scattering is represented by the *first* term in this series, so it is certainly a special (or limiting) case of Mie theory! For larger particles, more terms have to be included; indeed, for rainbows (produced by light scattering from raindrops—see the next chapter) the number of terms required is of the order of *five thousand*—the ratio of the drop circumference to the wavelength!

By now the wide-awake reader may be thinking: *So, why isn't the sky violet, since that has a shorter wavelength than blue light?* The reasons depend on both external and internal factors; firstly, sunlight is not uniformly intense at all wavelengths (otherwise it would be pure white before entering the atmosphere). It has a peak intensity somewhere in the green part of the visible spectrum, so the entering intensity of violet light is considerably less than that of blue. The other reason is physiological in origin: our eyes are less sensitive to violets than blues. The scattering of sunlight by molecules and dust is much more complex than described here, of course; there are subtle dependences of color and intensity as a function of angle from the sun and polarization of the light, and dust is typically *not* small compared to the the wavelengths of light, so Rayleigh scattering does not occur. In other words, it gets a lot more complicated! Excellent readable and informative accounts may be found in the books by Bohren, by Hoeppe, by Lynch and Livingston, by Naylor, and by Pesic, listed in the references.

Q.46: **So how much more *is* violet light scattered than red?** [color plate]

This question is really just an excuse to apply the Rayleigh's inverse fourth power law of scattering mentioned above. That's all we need to know here; since the question involves a comparison (and hence ratio) of scattering effects, no constants of proportionality enter our calculation—which is probably just as well because I'd have to look them up! When I was a student the common measure for wavelengths in the visible spectrum was the angstrom

(Å); 1 Å being 10^{-10} m. Nowadays, the nanometer (1 nm $= 10^{-9}$ m) is preferred. To keep the arithmetic simple, we choose the extremes of red light with wavelength $\lambda_r = 700$ nm (7000 Å) and violet light of wavelength $\lambda_v = 400$ nm. Clearly, the ratio of scattered violet light to scattered red light is given by

$$\frac{S_v}{S_r} = \left(\frac{\lambda_v}{\lambda_r}\right)^{-4} = \left(\frac{7}{4}\right)^4 = \frac{2401}{256} \approx 9.4.$$

We can think of this as a type of "direct" problem. A simple but illustrative inverse problem might be: if red light of wavelength 7000 Å, i.e., 700 nm, causes a certain amount of Rayleigh scattering, light of what wavelength will cause exactly three times as much scattering? Again, from the Rayleigh law, the unknown wavelength λ will be determined from

$$\frac{S_\lambda}{S_r} = 3 = \left(\frac{\lambda}{\lambda_r}\right)^{-4} \Rightarrow \lambda = \lambda_r \left(\frac{S_\lambda}{S_r}\right)^{-1/4} = 700 \times (3)^{-1/4} \text{ nm} \approx 532 \text{ nm},$$

i.e., in the green part of the spectrum. The first problem pretty much spans the visible part of the electromagnetic spectrum, and so provides us with a look at just how important the concept of scattering is in the world around us.

Q.47: **What causes variation in colors of butterfly wings, bird plumage, and oil slicks?** [color plate]

Who has not spent at least a moment or two enjoying the shimmering colors on the surface of a soap bubble as it floats by, the subtle changes of color in the throat plumage of a hummingbird, or the brilliant "eyes" in the tail feathers of a peacock? What about the feathers on the neck of a mallard drake, or the wings of a tropical butterfly? Oil slicks can also exhibit beautiful colors, which change in time as the oily film thins. Regardless of the context, all these shifting, shimmering colors arise from the fact that light is a wave, and waves can interact and *interfere* in some very interesting ways. As we will see, such interference occurs when the reflecting layers are about a wavelength or two apart. The colors that are observed as a result of interference are very sensitive to the film thickness, which is why we see so many colors on the surface of an ever-changing soap film surface, or on a thinning slick of oil or puddle of gasoline as we slightly change our direction of viewing.

A similar effect is caused by the scales on some butterfly wings; they are made of *chitin*, which is a horny substance also found in the shells of beetles

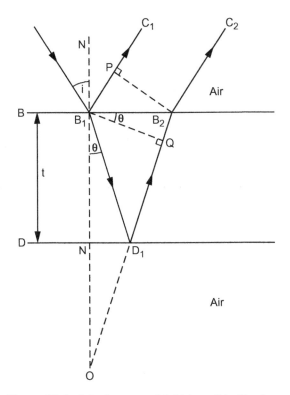

Figure 47.1. Interference of light in a thin film layer

and exoskeletons of insects. Reflections from these layers of chitin give rise to
varying colors as we examine carefully the wings. *Blue morpho butterflies*, na-
tive to Central and South America, have wings that are so bright that on occasion naturalists have reported seeing flashes of blue hundreds of yard away.
The layers of chitin in their wings are about a thousand times thinner than a
human hair, that is, about 10^{-7} m (or 100 nm), and this is about one-fourth
the wavelength of blue light. The layers are closely packed scales that overlap
like shingles or tiles on a roof. Jewel beetles, black-winged damselflies, and
peacock tails also exhibit iridescence, as does iron pyrite (a compound of iron
and sulfur), which has a thin layer of iron oxide on its surface, permitting
light interference to occur. Iridescence is also produced by alternating layers
of calcium carbonate and water in abalone shells; these are large, rather flat
shells lined with mother-of-pearl.

Consider in figure 47.1 light of monochromatic wavelength λ incident at
an angle of incidence i on a film of uniform thickness t. Some of the light will
be reflected and some transmitted (and subsequently reflected, etc; see figure). The angle of refraction is θ. The interference pattern will be determined

by the two parallel rays B_1C_1 and B_2C_2 emerging from the upper surface B. The refractive index of air will be assumed to be unity, and that of the film is $n > 1$. We consider the optical path difference Γ between light rays arriving by direct reflection (in direction B_1C_1) and that emerging by refraction and reflection along the path $B_1D_1B_2C_2$. Because the speed of light c, the wavelength λ, and the (fixed) frequency v are related by the equation $c = \lambda v$, it follows that within the soap film (or whatever the refracting medium happens to be) the wavelength is decreased because the speed of light in the medium is decreased. This means that the number of wavelengths per unit length in the film has increased, so in anthropomorphic terms, the ray "sees" a greater distance to travel within the film than compared with an equal thickness in air. This effective distance is increased by a factor n. Thus, from the geometry of the problem

$$\begin{aligned}
\Gamma &= n\,(B_1D_1 + D_1B_2) - B_1P \\
&= n\,(OD_1 + D_1B_2) - B_1P \\
&= n\,(OQ + QB_2) - B_1P.
\end{aligned} \tag{47.1}$$

It is clear that $QB_2 = B_1B_2 \sin\theta$ and $B_1P = B_1B_2 \sin i$ so that

$$n = \frac{\sin i}{\sin\theta} = \frac{B_1P}{QB_2}, \tag{47.2}$$

or $n \times QB_2 = B_1P$, so that the path difference between C_1 and C_2 is given by

$$\begin{aligned}
\Gamma &= nOQ = n\,(2NB_1 \cos\theta) \\
&= 2nt \cos\theta.
\end{aligned} \tag{47.3}$$

This is the fundamental relationship we require, and before proceeding further some comments are in order. At normal incidence ($i = 0$), $\theta = 0$ by Snell's law, and $\Gamma = 2nt$, which is just twice the optical thickness of the film. In addition, there is a phase change of π radians (or half a wavelength) that occurs when light is reflected at an air/medium boundary (meaning that the ray in air encounters the boundary with the denser medium); this does not occur at a medium/air boundary. Because of this there is an "effective" path difference between the two rays given by

$$\tilde{\Gamma} = 2nt \cos\theta + \frac{\lambda}{2}. \tag{47.4}$$

The two rays will interfere *constructively* if $\tilde{\Gamma} = N\lambda$, i.e.,

$$2nt \cos\theta = \left(N - \tfrac{1}{2}\right)\lambda, \qquad (47.5)$$

where $N = 1, 2, 3, \ldots$, because maxima will coincide with maxima, etc., and so this will correspond to a maximum of the interference pattern. Similarly, a minimum of the pattern will occur as a result of *destructive* interference when

$$2nt \cos\theta = N\lambda. \qquad (47.6)$$

Since any path difference corresponds to a difference in phase δ we shall calculate that by considering the rays emerging at C_1 and C_2 to have amplitudes A and $A_r = A(1 + e^{i\delta})$, respectively, where we may take A as a real number without loss of generality. The phase difference is

$$\delta = \frac{2\pi}{\lambda}\tilde{\Gamma} = \frac{2\pi}{\lambda}\left(2nt\cos\theta + \frac{\lambda}{2}\right). \qquad (47.7)$$

The total emerging intensity associated with these two rays is $I_r = |A_r|^2 = A_r A_r^*$, where the asterisk denotes complex conjugation, so

$$
\begin{aligned}
I_r &= A^2(1 + e^{i\delta})(1 + e^{-i\delta}) = A^2(2 + e^{i\delta} + e^{-i\delta}) \\
&= 2A^2(1 + \cos\delta) = 4A^2\cos^2\frac{\delta}{2} = 4A^2\cos^2\left(\frac{2\pi}{\lambda}nt\cos\theta + \frac{\pi}{2}\right) \\
&= 4A^2\sin^2\left(\frac{2\pi}{\lambda}nt\cos\theta\right).
\end{aligned}
\qquad (47.8)
$$

Obviously, each wavelength will have a corresponding value of I_r. For *extremely* thin films, i.e., when $t/\lambda \ll 1$, we have $I_r \approx 0$; this means that the "color" we see emerging is because the two reflected waves are almost exactly out of phase (there is zero reflected intensity). If the thickness of the film is increased, then the shorter wavelengths will be the first ones to interfere constructively, and give bright violet interference fringes, followed by blue, green, and so on. An interesting discussion of these colors in a variety of different situations (including umbrella fabric and window ice!) is given by Minnaert. There are different *orders* of interference colors associated with the integers N, and in a draining soap film (for example) it is possible to see several at once.

And a closely related question follows . . .

Q.48:   What causes the metallic colors in that cloud?
[color plate]

A good question. Such cloud appearances can be exquisitely beautiful, and very transient. The phenomenon is often referred to as *iridescence* or *irisation*. Essentially, the culprit is *diffraction*: the bending of waves around objects. Diffraction can occur for all types of waves. Water waves can be diffracted by obstacles like piers and harbor features. Perhaps you have noticed, in an auditorium of some kind or a church, in which your view of a speaker is blocked by a pillar, that you can still *hear* what is being said? Why can your ears receive auditory signals, but your eyes cannot receive visual ones (excluding Superman for the moment)? The reason for this is related to the lengths of the sound and light waves, being $\simeq 1$ m and $\simeq 10^{-7}$ m, respectively. The latter, in effect, *scatter* more like a particle, while the former are able to *diffract* (bend) around an obstacle comparable in size to its wavelength. By the same token, therefore, we would expect that light waves can diffract around appropriately smaller obstacles, and indeed this is the case, as evidenced by softly colored rings of light around the moon (coronae; see question 91) as thin clouds scud past its face. The preceding question dealing with iridescence on bird plumage, butterfly wings, or oil slicks is closely related to this one because, interference of light is closely related to diffraction. In short, interference of waves gives rise to diffraction fringes, though physically there is little difference between the phenomena, except that traditionally interference seems to be the preferred term to use when the number of interacting waves is small, and diffraction is used to describe the effects of many such waves interacting.

Obviously, for this question, the cloud droplets are the obstacles encountered by the sunlight; and the longer the wavelength the greater the amount of diffraction, so red light is bent more than violet. Because of this differential diffraction effect, constructive and destructive interference of light ensures that color bands result, and this effect is most prominent near the edge of the cloud, where it is thin, and the cloud droplets, being new, may be more uniform in size. In other parts of the cloud this effect is reduced because of nonuniformity of drop size and also multiple scattering from cloud droplets. When the drops have very uniform size, the colors tend to be brighter and purer. If they have varying sizes, the colors are more washed out and whitish. The phenomenon is essentially the same as that producing a corona; iridescent clouds are essentially corona fragments, so we will discuss a little bit of the underlying mathematics here, and refer to it in a little more detail in question 91, concerning a lunar corona.

The intensity of the light of wavelength λ that is diffracted, or bent, through an angle θ by a cloud droplet of radius R is a function of the quantity

$$x = \frac{2\pi R}{\lambda}\sin\theta,\qquad(48.1)$$

i.e., to the ratio of the drop circumference to the wavelength, modulated by $\sin\theta$. In fact, the function of x is itself oscillatory, being a Bessel function (see question 91 and appendix 1), but that does not concern us here. As $|x|$ increases from zero, the intensity of diffracted light falls steeply from its maximum value to zero, followed by a series of oscillations decreasing in amplitude (see figure 91.1). Therefore, sufficiently small values of $|x|$ (i.e., less than the location of the first zero) correspond to higher intensity of light. Without going into unnecessary mathematical detail here, we can glean some useful information from the expression (48.1). Note that for given angle θ and wavelength λ, smaller values of $|x|$ correspond to smaller values of R, i.e., smaller cloud droplets. For given values of R and θ, larger wavelengths (e.g., reddish light) correspond to smaller values of $|x|$, and therefore higher intensities. For given values of R and λ, smaller values of θ in the interval $(0, \pi/2)$ also correspond to higher intensities. Of course, in a cloud, drop sizes may vary considerably, and the intensity of different colored iridescent fragments depends quite sensitively on angle and drop size.

Q.49: How do rainbows form? And what are those fringes underneath the primary bow? [color plate]

FORMATION

What *is* a rainbow? A rainbow is sunlight displaced by reflection and dispersed by refraction in raindrops, seen by an observer with his or her back to the Sun (under appropriate circumstances). The *primary* rainbow, which is the lowest and brightest of two that may be seen, is formed from two refractions and one reflection in myriads of raindrops (see figure 49.1). It can be seen and photographed, but it is not located at a specific place, only in a particular direction. Obviously, the raindrops causing it are located in a specific region in front of the observer. The path for the *secondary* rainbow is similar, but involves one more internal reflection. In principle, an unlimited number of higher-order rainbows exist from a single drop, but light loss at

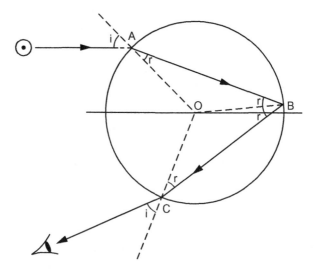

Figure 49.1. Ray path inside a spherical raindrop for the primary bow

each reflection limits the number of rainbows to two (claims have been made concerning observations of the *tertiary* bow, but such a bow would occur around the Sun, and be very difficult to observe, quite apart from its intrinsic faintness.) While each drop produces its individual primary rainbow, what is seen by an observer is the cumulative set of images from myriads of drops; some contribute to the red region of the bow, while others contribute the orange, yellow, green, and so forth. Although each drop is falling, there are numerous drops to replace each one as it falls from a particular location, and so the rainbow, for the period that it lasts, is for each observer effectively a continuum of colors produced by a near-continuum of drops.

Let's start with an examination of the basic geometry for a light ray entering a spherical droplet. From figure 49.1 note that after two refractions and one reflection the light ray shown contributing to the rainbow has undergone a total deviation of $D(i)$ radians, where

$$D(i) = \pi + 2i - 4r, \tag{49.1}$$

in terms of the angles of incidence (i) and refraction (r), respectively. The latter is a function of the former, this relationship being expressed in terms of Snell's law of refraction

$$\sin i = n \sin r, \tag{49.2}$$

where n is the relative index of refraction (of water, in this case). This relative index is defined as

$$n = \frac{\text{speed of light in medium I (air)}}{\text{speed of light in medium II (water)}} > 1. \qquad (49.3)$$

Since the speed of light in air is almost that in vacuo, we will refer to n for simplicity as *the* refractive index; its generic value for water is $n \approx 4/3$, but it does depend slightly on wavelength (this is the phenomenon of dispersion, and without it we would have only bright "whitebows"!)

Let's now see how $D(i)$ behaves. Note that $D(0) = \pi$. Firstly, a little differentiation is in order. Thus,

$$\frac{dD}{di} = 2 - 4\frac{dr}{di}, \qquad (49.4)$$

and differentiating Snell's law we find that

$$\cos i = n \cos r \frac{dr}{di}, \qquad (49.5)$$

so that

$$\frac{dD}{di} = 2 - \frac{4\cos i}{n\cos r}. \qquad (49.6)$$

Are there any critical numbers in the domain $i \in [0, \pi/2]$? If so, this would mean that D is stationary for those values of i (at i_c, say). Physically, this would correspond to a concentration of deviated rays in the angular region about $D(i_c)$. Let's see: $D'(i) = 0$ implies that

$$\frac{1}{4} = \frac{\cos^2 i}{n^2 - 1 + \cos^2 i}, \qquad (49.7)$$

upon rearranging things and using Snell's law again. There are no spurious solutions that have been introduced by squaring quantities, so at an extremum it follows that

$$\cos i = \left(\frac{n^2 - 1}{3}\right)^{1/2} \equiv \cos i_c. \qquad (49.8)$$

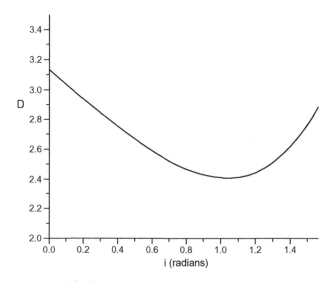

Figure 49.2. The deviation angle $D(i)$ defined by (49.9)

It can be shown that $D''(i_c) > 0$ so that the deflection $D(i_c)$ is a minimum. (The phrase *it can be shown that* has annoyed and frustrated generations of students, myself included; nevertheless, my only response at this time is an equally annoying one—this is left as an exercise for the interested reader). Note that $D(i)$ may be written with sole dependence on i as

$$D(i) = \pi + 2i - 4\arcsin\left(\frac{\sin i}{n}\right), \tag{49.9}$$

and this is illustrated in figure 49.2 for the case of an value for water of $n = 4/3$ (the axes are in radians).

For a "generic" monochromatic rainbow (the whitebow referred to above), the choice $n = 4/3$ yields

$$i_c = \arccos\sqrt{\frac{7}{27}} \approx 1.04 \text{ radians} \approx 59.4°,$$

and $D(i_c) \approx 2.41$ radians $\approx 138°$. The supplement of this angle in degrees, $180° - D(i_c) \approx 42°$ is the semi-angle of the rainbow "cone" formed with apex at the observer's eye, the axis being along the line joining the eye to the anti-solar point.

Now for a little comment about the *color dispersion* in a rainbow. While it is not standardized to the satisfaction of everyone (as far as I can tell), the visible part of the electromagnetic spectrum extends from the red end (700–647 nm) to the violet end (424–400 nm), a *nanometer* (nm) being 10^{-9} m. For red light of wavelength $\lambda \approx 656$ nm, the refractive index $n \approx 1.3318$, whereas for violet light of wavelength $\lambda \approx 405$ nm, the refractive index $n \approx 1.3435$—a slight but very significant difference! All that has to be done is the calculation of i_c and $D(i_c)$ for these two extremes of the visible spectrum, and the difference computed. Then, voilà! We have the angular width of the primary rainbow. In fact, since $D(i_c) \approx 137.75°$ and $139.42°$ for the red and violet ends, respectively, the angular width $\Delta D \approx 1.67° \approx 1.7°$, or about three full moon angular widths.

WHAT THOSE FRINGES ARE: SUPERNUMERARY BOWS

An important point must be made concerning the level of mathematical sophistication involved in this model of the rainbow. We have been using what is often referred to as *geometrical optics*; this is the branch of optics that treats the interactions of light with objects in terms of rays, with no regard to the wave nature of light. As such, while geometrical optics enables us to predict both the angular distribution of the rainbow colors and their apparent location relative to the observer and the sun, it cannot predict or explain the existence of the *supernumerary bows*. These are sometimes visible as several pastel colored fringes below violet part of the primary bow. In principle, they can also exist above the secondary bow, but they are rarely, if ever observed. Newton, being an advocate of the corpuscular theory of light, was unable to explain the existence of these supernumerary (literally, *superfluous*) bows.

They are the result of interference of light waves. Thus, two rays that enter the drop on either side of the rainbow ray (the ray of minimum deviation) may exit the drop in parallel paths; this will happen for appropriately incident rays. By considering the wavefronts (perpendicular to the rays), the incident waves will be in phase (i.e., crests and troughs aligned with crests and troughs). But inside the drop they travel paths of different length. Depending on whether this path difference is an integral number of wavelengths or an odd integral number of half-wavelengths, these waves will reinforce each other (constructive interference) or cancel out each other (destructive interference). Obviously, partially constructive/destructive interference can occur if the path difference does not meet the above criteria. Where waves reinforce one another, the intensity of light will be enhanced; conversely, where they annihilate one another the intensity will be reduced. Since these beams of light will exit the raindrop at a smaller angle to the axis than the "rainbow"

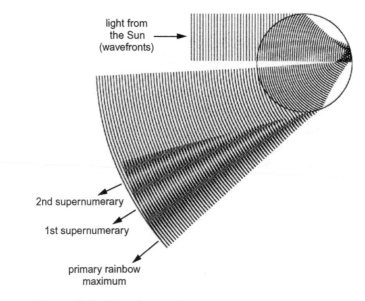

light from
the Sun ⟶
(wavefronts)

2nd supernumerary

1st supernumerary

primary rainbow
maximum

Figure 49.3. Wavefront refraction and supernumerary bows

ray, the net effect for an observer looking in this general direction will be a series
of light and dark bands just inside the primary bow. They are just as much a
part of the "rainbow" as the primary bow.

Actually, even the above explanation is not strictly true; the limitation
comes from thinking in terms of the interference of *two waves*. It is more accurate to regard the supernumerary bows (and the primary bow) as resulting
from different parts of the *same* wave interfering, resulting in the sort of moiré
pattern shown in figure 49.3.

Supernumerary bows rarely extend around the fully visible arc, for reasons
that are related to drop size. The angular spacing of these bands depends
on the size of the droplets producing them. The width of individual bands
and the spacing between them decreases as the drops get larger. If drops of
many different sizes are present, these supernumerary arcs tend to overlap
somewhat and smear out what would have been obvious interference bands
from droplets of uniform size. This is why these pale blue or pink or green
bands are most noticeable near the top of the rainbow: it is the smaller drops
that contribute to this part of the bow, and these may represent a rather narrow range of sizes. Nearer the horizon a wide range of drop size contributes
to the bow, but this tends to blur the interference bands.

Although geometrical optics is an important and useful branch of optics,
another deficiency in connection with regions of "light and dark" is that it predicts that the rainbow has infinite intensity. Clearly, this is a result of the

Question 24. A cylindrical "hay-stack" roll; the calculations for a hemispherical haystack are readily adapted for this geometry (and for the roll lying on its side). The conclusions are very similar, apart from geometric factors.

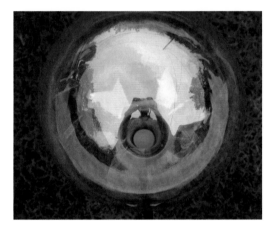

Question 25. The "universe" produced by reflection from a garden globe. The author cannot avoid being part of the view!

Question 45. Part of a cumulus congestus cloud seen against the blue sky.

Question 46. A striking sunrise viewed from outside the author's home in Nor-folk, Virginia. The red and orange colors are present because at sunrise (and sunset) the sunlight traverses a relatively long path through the atmosphere, and most of the shorter wavelengths (violet and blue) have been scattered out of the line of sight.

Question 47. A blue morpho butterfly. The rich iridescent colors are caused by the interference of light waves. Courtesy of Barbara Tierney.

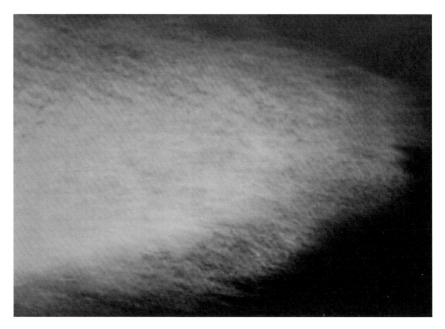

Question 48. Iridescence in a cloud located about 30^0 from the sun. This delicate phenomenon is produced by the diffraction of sunlight from cloud droplets.

Question 49. A (primary) spraybow seen at the Lower Falls in Yellowstone National Park.

Question 50. Primary and secondary spraybows seen at the Upper Falls in Yellowstone National Park.

Question 54. A multitude of bubbles!

Question 57. A sun pillar, here formed above the setting Sun, and created by rays reflected downward from the lower faces of myriads of tilted plate-like ice crystals. The larger the tilt, the longer is the pillar.

Question 58. A sundog (or parhelion, or mock sun) located at the same altitude as the Sun. Sundogs can be found at least 22^0 away from the Sun (depending on the solar altitude), and may be seen on either or both sides of the Sun, depending on the location of the cirrostratus clouds producing them. Like the 22^0 halo, they are formed by sunlight being refracted through the sides of hexagonal plate-like ice crystals in cirrostratus (or cirrus) clouds, but unlike that halo, these plates drift and float gently downward with their large hexagonal faces almost horizontal.

Question 60. A glory, photographed not long after take-off from London's Gatwick airport. The plane was close enough to the cloud bank that its shadow was quite evident. Glories are formed when light is backscattered by individual cloud droplets, though the contributing mechanisms are not all straightforward ones.

Question 68. A glitter path. Photograph taken from the Leif Erikson ferry, looking across to Nova Scotia, courtesy Heather Renyck.

Question 82. The edge of a wood in Banff National Park, Alberta.

Question 85. Shadow of Everest at sunrise. Courtesy Wally Berg, Berg Adventures International (www.bergadventures.com).

Question 86. Zion Arch, Zion National Park, Utah.

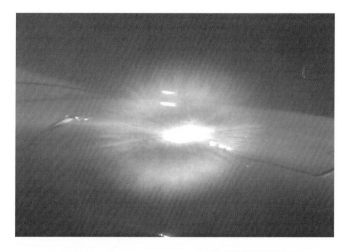

Question 91. An elliptical corona seen on a flight from Calgary to Houston as we descended. It was caused by several things. The picture shows the reflection of the Sun from the aircraft engine housing on the wing, viewed through a layer of tiny water droplets on one of the inner window surfaces! The reflection of the Sun appeared elliptical because of the curvature of the metallic surface, and the light from that source was diffracted by the tiny droplets (causing the observed interference pattern), resulting in this elliptical spectrum of colors around the source.

inadequacy of the model at such boundaries, or anyone who looked at a rainbow would be blinded! A discussion of this particular issue—a singularity implied by geometrical optics—is deferred until question 61 at the end of this section.

Finally, the table 49.1, adapted from the book by Minnaert, is a qualitative description of rainbow colors as an approximate guide to the size of raindrops producing them.

TABLE 49.1
Variation of Rainbow Colors with Drop Size

Drop diameter (mm)	Description of rainbow colors
1–2	Very bright violet and vivid green; the bow contains pure red, but scarcely any blue. Supernumerary bows are numerous (there may be as many as five, or even more), violet-pink alternating with green, and merging continuously into the primary bow.
0.5	The red color is considerably weaker, and there are fewer supernumerary bows, again with alternating violet-pink and green fringes.
0.2–0.3	No red is apparent; the bow is broad and well developed. The supernumerary bows become more yellow. If a gap occurs between these bows, the diameter of the drops is no more than about 0.2 mm. If there is a gap visible between the primary bow and the first supernumerary bow, the diameter of the drops is less than 0.2 mm.
0.08–0.10	The bow is now broader and paler, with only the violet vivid. The first supernumerary bow is well separated from the primary bow by a fairly wide gap and white tints may be clearly identified.
0.06	The primary bow contains a distinct white stripe.
< 0.05	The bow is now a somewhat colorless cloud bow or fogbow.

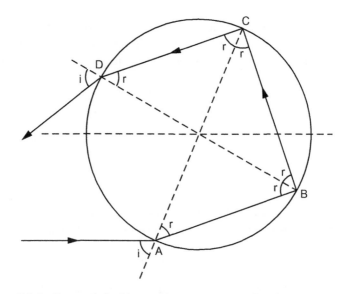

Figure 50.1. Ray path inside a spherical raindrop for the secondary bow

Q.50:   What about the secondary rainbow? [color plate]

This time there is an additional reflection, so the geometry of figure 50.1 yields
a total ray deviation of

$$D(i) = 2\pi + 2i - 6r = 2i - 6r \; (\text{modulo } 2\pi) = 2i - 6 \; \arcsin\left(\frac{\sin i}{n}\right). \quad (50.1)$$

Proceeding as before we find that

$$\frac{dD}{di} = 2 - 6\frac{dr}{di} = 2 - 6\frac{\cos i}{n \cos r}, \quad (50.2)$$

so that critical numbers occur when

$$\frac{1}{9} = \frac{\cos^2 i}{n^2(1 - \sin^2 r)} = \frac{\cos^2 i}{n^2 - \sin^2 i}, \quad (50.3)$$

or

$$\cos i = \left(\frac{n^2 - 1}{8}\right)^{1/2} \equiv \cos i_c. \quad (50.4)$$

For our generic n-value of $4/3$,

$$\cos i_c = \sqrt{\frac{7}{72}} \approx 0.3118, \text{ so } i_c \approx 71.8°,$$

so that

$$D(i_c) = 2i_c - 6\arcsin(0.7125) = 143.6° - 276.6° = -129°. \qquad (50.5)$$

The negative sign is of no significance here, since the ray geometry is cylindrically symmetric about the central axis, but for this reason sometimes $D(i)$ is defined as the negative of the expression used here. Just as the supplement of $D(i_c)$ for the primary bow is approximately 42°, so for secondary bow the supplementary angle is 51°; thus, the secondary bow is about 9° higher in the sky than the primary bow.

Q.51:   Are there higher-order rainbows?

For k internal reflections, continuing as we have done, it is a straightforward matter to show that

$$D_k(i) = k\pi + 2i - 2(k+1)\arcsin\left(\frac{\sin i}{n}\right), \qquad (51.1)$$

and after an interesting trigonometric calculation, described below, the following result can be derived for $k = 1$,

$$D_1(i_c) = 2\arccos\left[\frac{1}{n^2}\left(\frac{4-n^2}{3}\right)^{3/2}\right], \qquad (51.2)$$

thus expressing the angle of minimum deviation (the "rainbow angle") in terms of the refractive index alone. This was first derived by Huyghens in 1652. In principle, it can also be done for higher-order rainbows—though, to my knowledge, this has not been carried out before (see Adam 2008)—but the algebra gets messy (details are available on request). Substituting $n = 4/3$ into the above equation gives

$$D(i_c) = 2\arccos\left[\frac{9}{16}\left(\frac{20}{27}\right)^{3/2}\right] = 2\arccos(0.3586) \approx 138°. \qquad (51.3)$$

The above standard ray–theoretic calculus-based approach to rainbow formation in spherical droplets has been known since the time of Newton (indeed, it was much studied by him, although Descartes contributed greatly to the underlying ray theory). Figures 49.1 and 50.1 show the ray paths for $k = 1$ and 2. As noted above, the reason a rainbow exists at all (and would do even in the absence of spectral dispersion, as a "whitebow") is because there is an extremum—a minimum to be precise—in $D_k(i)$ at the critical angle of incidence i_c defined by the vanishing of $D'_k(i)$. Again, it is a standard exercise to establish that

$$ i_c = \arccos \left[\frac{n^2 - 1}{k(k+2)} \right]^{1/2} . \tag{51.4} $$

For a specified number of reflections k there is clearly a limit on the value of n (namely, $1 \leq n \leq k+1$), though in practice this is not an issue (on planet Earth, at least) since $n \approx 4/3$ for water. Of course, n is slightly wavelength dependent in the optical spectrum, and it is this that gives rise to the *colors* (and, indeed, the width) of the rainbow, though as we have seen, this is not a necessary condition for the existence of this curved "caustic" in the sky . . . (For $k = 1$ and $n = 4/3$, equation (51.4) yields $i_c \approx 59°$ as we have seen.) A further point to note is a major consequence of the theory of geometrical optics: this theory predicts that the intensity of light at the rainbow angle $D(i_c)$ is infinite! (See question 61 for details). Clearly, a more sophisticated theory—the wave theory of light—is needed to go beyond this very crude prediction. While outside the scope of this book, further references are provided in the bibliography.

A TRIPLE RAINBOW?

Frequently, someone will tell me that they saw a "triple rainbow," and per-haps something in my expression suggests skepticism, because they empha-size what a truly magnificent sight it was. And I have no doubt that it was. My skepticism arises from the implication that what was seen was another rainbow beyond the secondary, i.e., a tertiary rainbow. Even as I write this, I realize that my skepticism is ill-founded, because most people aren't even aware of the extreme improbability of seeing such a bow, and, instead, they must be describing something they see close to the primary and secondary bows. The reasons why a true tertiary bow is so unlikely to be seen are dis-cussed in the very next question. Returning to the above conversation, the next comment I make is to ask my friend if there was a body of water either behind them or in front of them when they saw the "triple bow." Frequently, the response is, "I don't know," so the problem remains unresolved, at least until (in my mind) we examine question 52 below.

Q.52:   So what *is* that triple rainbow?

In principle, as already noted, more than two internal reflections may take place inside each raindrop, so higher-order rainbows (tertiary, quaternary, etc.) are possible. It is possible to derive the angular size of such a rainbow after any given number of reflections (Newton was the first to do this). Newton's contemporary Edmund Halley found that the third rainbow arc should appear as a circle of angular radius about 42° around the Sun itself! The fact that the sky background is so bright in this vicinity, coupled with the intrinsic faintness of the bow itself would make such a bow difficult, if not impossible, to see (but see the claim by Pedgely in the references). However, in the carefully controlled conditions of a laboratory, the situation is very different. Jearl Walker has used a laser beam to illuminate a single drop of water and traced rainbows up to the thirteenth order, their positions agreeing closely with predictions. Others have traced nineteen rainbows under similar laboratory conditions.

Let's "do the math" ourselves for the tertiary bow. From equation (51.4) with $k = 3$ and $n = 4/3$, the critical angle of incidence for the tertiary bow is

$$i_c = \arccos\left(\frac{7}{135}\right)^{1/2} \approx 1.34 \text{ radians} \approx 76.8°. \qquad (52.1)$$

Then using (51.1) we find that the corresponding ray is deviated through an angle of

$$D_k(i_c) = 3\pi + 2i_c - 6\arcsin\left(\frac{3\sin i_c}{4}\right) \approx (-)41.6°. \qquad (52.2)$$

The negative sign is once again immaterial here, because of the cylindrical symmetry of the problem. Coincidentally though, the angular radius of this bow is close to that of the primary, but it's in the opposite direction! As noted above, it would appear as a very faint ring around the Sun with this angular diameter, *if* it could be seen above the glare of the sun in that part of the sky, which is highly unlikely. As we will see in question 56, frequently rings do occur around the Sun, but they are caused by ice crystals, not raindrops!

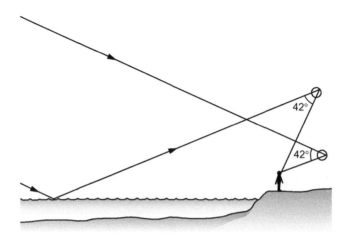

Figure 52.1. Ray paths for regular and reflected-light primary bows (redrawn from Greenler (1980))

So what about these anomalous "triple rainbows" that people have seen? Recall my question about the presence of a nearby body of water? If the water is behind the observer, i.e., between the observer and the Sun, he or she may see a reflected light or *reflection* rainbow. Sunlight reflected from the water surface behind the observer will then be scattered by a set of raindrops different from those causing the primary and secondary bows (see figure 52.1). This reflected rainbow will be higher in the sky than the primary, but will meet it at the horizon, as shown below. It too is a "primary" bow (though, in principle, a still higher secondary may also be seen), so the colors are in the same order, red being on the outside of the bow. It will probably be fainter than the "primary" primary, water not being a perfect reflector. When seen together, three or possibly *four* bows might be seen—a truly magnificent sight, on occasion, I am sure. More commonly (and especially if the body of water is not particularly large), only a lower small segment of this bow is seen, and will appear as an almost vertical appearing to emanate from the foot of the primary bow. This may be the most common source of "triple bow" reports.

Below we show using geometry and algebra that a rainbow and its reflected-light counterpart intersect at the horizon no matter what the elevation of the sun. This result is true for both primaries and secondaries (so we need not establish the result twice). We assume, without loss of generality for this problem, a colorless rainbow of zero thickness, i.e., an arc of a circle, as in figure 52.2.

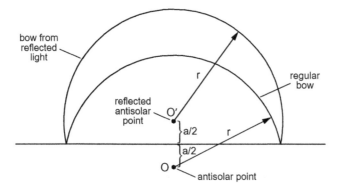

Figure 52.2. The regular and reflected-light primary rainbows; a body of water is behind the observer (redrawn from Greenler (1980))

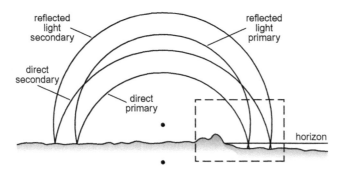

Figure 52.3. Configuration for the existence of multiple rainbow "shafts" (redrawn from Greenler (1980))

(i) Either we can make an appeal to symmetry—the graph of two identical partially overlapping circles is symmetric about the line joining their points of intersection, or (ii) we can let O at $(0, 0)$ be the center of the the lower circle of radius r, so that if the center O' of the reflected bow is displaced by a distance a (where a can be of any sign, but in this case it will be positive), the equations of the two circles are $x^2 + y^2 = r^2$ and $x^2 + (y - a)^2 = r^2$. At the points of intersection, $a\,(2y - a) = 0$, the nontrivial result being that these points lie on the line of symmetry, i.e., the horizon, $y = a/2$. (The centers of the rainbow and the reflected bow are symmetrically situated above and below the horizon.)

In principle, up to eight rainbow "shafts" could be observed under the right circumstances, with both primary and secondary regular (direct) and reflected light bows, as illustrated in figure 52.3.

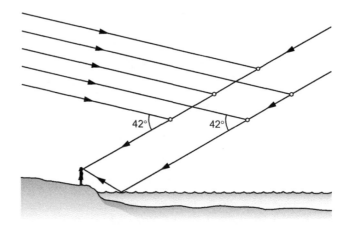

Figure 52.4. Ray paths for the regular and reflected primary bows (redrawn from Greenler (1980))

We turn finally to the case of a *reflected* rainbow (see figure 52.4). Now the body of water must be in front of the observer, but, ironically, the reflection is *not* of a rainbow the observer will see in the sky! Light from this "invisible" rainbow is reflected from the water surface into the observer's eye, and the resulting image is, of course, an inverted rainbow, entirely separate from the visible primary bow, so the "feet" do not coincide as they would if the former were a true reflection of the latter.

Q.53:   Is there a "zeroth-order" rainbow?

That is to say, can rays impinge upon a raindrop, be refracted into the drop, and then be refracted out into the air directly, forming a rainbow? (See figure 53.1.) Such a process can occur, but no rainbow will result. As can be seen from the diagram, the deviation angle is $D_0(i) = 2\,(i-r)$, so using Snell's law of refraction,

$$D_0'(i) = 0 \Longleftrightarrow dr/di = 1 \Longrightarrow \cos i = n\cos r, \qquad (53.1)$$

but this implies that

$$\cos^2 i = n^2 - \sin^2 i \Longrightarrow n^2 = 1.$$

Mathematically, $n > 1$ implies a contradiction, and, physically, $n = 1$ implies there is no refraction; either way, there is no rainbow!

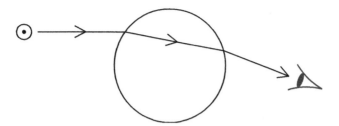

Figure 53.1. A direct ray path through a drop

Q.54:   Can bubbles produce "rainbows"? [color plate]

Is there a "rainbow" angle (and corresponding angle of incidence) for air bubbles? That is to say, can rainbows (or *bubblebows*) form from the scattering of light in water from air bubbles? (See figure 54.1.) We'll assume that the refractive index for water/air interaction is $n = 3/4$. We also need to note that *total internal reflection* can occur for light in a dense medium entering a less dense medium.

The simplest possibility involves rays entering the bubble, being refracted into the bubble, and then being refracted out into the water. This would correspond to the zeroth bow in the question above, i.e., with two refractions and *no* reflections (corresponding to deviated but direct transmission through a raindrop). If this gives rise to a rainbow, a secondary bow might occur (if the zeroth bow occurs we will call it the primary bow!) as a result of an additional interior reflection (see figure 54.1).

For rays entering a less dense medium, $n = (4/3)^{-1} = 3/4$. Total internal reflection occurs when $\sin i \geq n$, or (here)

$$i \geq \arcsin (3/4) \approx 48.6° = i_c.$$

Conversely, for transmission to occur, $i < i_c = 48.6°$. Subject to this constraint, the deviation angle $D_1(i) = 2 (r - i)$; from Snell's law of refraction

$$\sin i = n \sin r \text{ (and noting that } dr/di > 0) \implies dD/di = 2 \left(\frac{\cos i}{n \cos r} - 1 \right)$$

$$= 0 \text{ if } \cos i = n \cos r.$$

$$(54.1)$$

But a little algebra (as in question 53) shows that this implies n^2 must equal 1, which cannot be true for air bubbles in water, for which $n < 1$ in our problem. *Therefore, there is no bubblebow for this configuration.* What of a

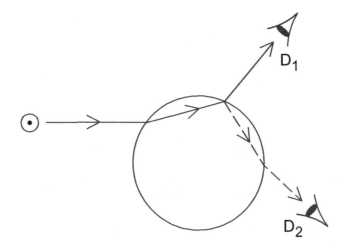

Figure 54.1. Ray paths through a bubble

possible secondary bow? Regarding the next possibility (corresponding to the primary bow in a raindrop), the deviation for two refractions and one reflection is

$$D_2(i) = 2(r - i) + (\pi - 2r) = \pi - 2i, \text{ so } D_2'(i) = -2, \qquad (54.2)$$

and there is no extremum at all; the deviation angle is a monotonically and, indeed, linearly decreasing function of i.

Q.55:   What would "diamondbows" look like?

Suppose that raindrops were made of diamonds instead of water (and who said mathematics was not a practical subject?). Diamond has a refractive index of $n \approx 2.42$ (for our generic "colorless" rainbow). Calculate the angles that a primary and secondary "diamondbow" would make with the antisolar direction, to the nearest tenth of a degree. We will assume the diamonds are spherical! Recall that for one internal reflection,

$$\cos i_c = [(n^2 - 1)/3]^{1/2}, \qquad (55.1)$$

i_c being the critical angle of incidence. Since θ is a real number, $|\cos \theta| \leq 1$, and it follows that a "primary" bow will arise from a single internal reflection only when $n^2 > 1$ and $n^2 - 1 \leq 3$, i.e., $1 < n \leq 2$. Clearly, this cannot occur for a diamond drop with the value of n as stated, so, as shown below, the diamondbow "primary" must occur for two internal reflections of the incident ray. The appropriate result is now that

$$\cos i_c = [(n^2 - 1)/8]^{1/2}, \tag{55.2}$$

and for $n = 2.42$, $i_c = \arccos (0.78) \approx 38.7°$, and hence the deviation

$$D(i_c) = 2i_c - 6\arcsin(\sin i_c/n) \approx -12.5°, \tag{55.3}$$

which corresponds to a small ring around the Sun with angular radius 12.5°. *Three* internal reflections will actually give rise to a secondary diamondbow, because now

$$\cos i_c = [(n^2 - 1)/15]^{1/2} \approx 55.3°, \tag{55.4}$$

and modulo 2π,

$$D(i_c) = \pi + 2i_c - 8\arcsin(\sin i_c/n) \approx 131.6°, \tag{55.5}$$

not too far from the value for the primary *rain*bow. Hmm . . . Lucy in the sky with diamonds, perhaps.

Q.56:   What causes that ring around the Sun?

Have you ever noticed a circular ring around the Sun (or, for that matter, the Moon) when the sky is clear except for wispy thin cirrus or cirrostratus clouds in the vicinity of the Sun? I read on a now-defunct website that such beautiful displays, known as *ice crystal halos,* can be seen on average twice a week in Europe and parts of the United States, and certainly my own experience is not terribly different, though I would estimate that I notice them about twice a month on average. The "radius" of a ring around the Sun or Moon is naturally expressed in terms of degrees of arc, subtended at the observer's eye by the apparent radius. The most frequently visible display is the the 22° circular halo, according to that site, followed by *parhelia* (or sundogs; see question 58) and then what is called an *upper tangent arc.* In my own rather limited experience, the sundogs are the most common, at least as I walk to work in southeastern Virginia. These are found at the same altitude as the Sun, close to (but just beyond) the 22° halo. A convenient and literal rule of thumb is that the outstretched hand at arm's length subtends about the same angle, so that if one's thumb covers the Sun, one's little finger extends to about the

Question 56. A 22° ice crystal halo, formed by sunlight being refracted through the sides of myriads of hexagonal plate-like ice crystals in cirrostratus clouds. In color, a reddish tinge can be seen on the inside boundary of the halo.

22° halo (*or* the sundog). One encyclopedic and very reputable website is that by Les Cowley, and I would strongly recommend that the reader consult this site (see the references for details).

Halos are formed (and seen) when sunlight is refracted, reflected, or both from ice crystals in the upper atmosphere and enters the eye of a *careful* observer: careful, because even the common types are easily missed. As already noted, they are usually produced when a thin uniform layer of cirrus or cirrostratus cloud covers large portions of the sky, especially in the vicinity of the Sun. Surprisingly, perhaps, they may occur at any time of the year, even during high summer, because above an altitude of about 10 km it is always cold enough for ice crystals to form. In particularly cold climes, of course, such crystals can form at ground level (though we are not thinking here of snow crystals). Very many of these crystals are hexagonal prisms; some are thin flat plates, while others are long columns, and sometimes the latter have bullet-like or pencil-like ends. A significant feature of all these crystals is that while any given type may have a range of sizes, the angles between the faces are the same. Although they do not possess *perfect* hexagonal symmetry, of course, they are sufficiently close to this that simple geometry based on such idealized forms suffices to describe many of the different arcs and halos that are associated with them. The halo formation that results from cirrus cloud crystals depends on two major factors: their shape and their orientation as

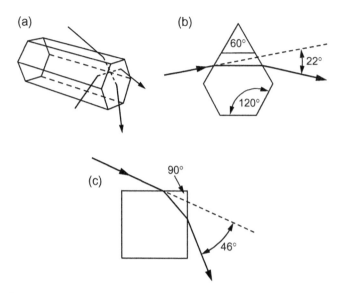

Figure 56.1. Hexagonal ice crystal and symmetric ray paths through them for the 22° and 46° halos

they fall. Their shape is determined to a great extent by their history, i.e., the temperature of the regions through which they drift as they are drawn down by gravity and buffeted around by winds and convection currents.

To explain the reasons for the various angles, e.g., 22°, it is necessary to examine the crystal geometry in more detail. It is clear from figure 56.1 that hexagonal ice crystals can be thought of as presenting "prism angles" of 60°, 90°, or 120° to rays entering them in the planes indicated, depending on the orientation of the crystal. There are both similarities and differences with the formation of rainbows in raindrops. First, we state things a little more formally in terms of theorems (stated here without proof; see Adam (2006) for further details) concerning the refraction of light through prisms of apex angle γ and (relative to air) refractive index n.

Theorem 1: The deviation or deflection angle for light refracted through a prism is a minimum for symmetric ray paths.

Recall that a rainbow arises when the angle through which a ray is deviated on passing through the raindrop is an extremum. This theorem tells us that in the case of *prisms*, the corresponding extremum—a minimum, in fact—occurs when the ray path through the crystal is symmetric, and the second theorem defines the magnitude of this minimum deviation angle D_m, i.e., the *location* of the halo relative to the line joining the sun and the observer.

Theorem 2: The minimum deviation angle D_m for a prism satisfies the relation

$$n = \frac{\sin[\frac{1}{2}(\gamma + D_m)]}{\sin\frac{1}{2}\gamma}. \qquad (56.1)$$

From this result we may solve for D_m to obtain

$$D_m = 2\arcsin\left(n\sin\frac{\gamma}{2}\right) - \gamma. \qquad (56.2)$$

We can use this result to explain the occurrence of the $22°$ and the less common $46°$ halos. As shown in figure 56.1(b, c), there are three prism angles in a hexagonal ice crystal prism: $60°$ (light entering side 1 and exiting side 3); $90°$ (light entering a top or bottom face and exiting through a side), and $120°$ (light entering side 1 and being totally internally reflected by side 2). The first two of these create color by dispersion of sunlight; the last contributes nothing directly to a halo, at least of interest to us here. The refractive index of ice for yellow light is $n \approx 1.31$. For the apex angle $\gamma = 60°$,

$$D_m = 2\arcsin(1.31\sin 30°) - 60° = 21.8° \approx 22°,$$

and for $\gamma = 90°$,

$$D_m = 2\arcsin(1.31\sin 45°) - 90° = 45.7° \approx 46°.$$

We can make this look really complicated by showing that the general expression for the deviation of a ray incident at angle i in a prism with apex angle γ is

$$D(i; \gamma) = i - \gamma + \arcsin\left\{n\sin\left[\gamma - \arcsin\left(\frac{\sin i}{n}\right)\right]\right\}. \qquad (56.3)$$

All of this goes to show, of course, that not only are ice crystals responsible for some magnificent displays in the sky from time to time, but also they are associated with some rather impressive-looking transcendental equations! Sketches of $D(i)$ are shown in figure 56.2 for the two apex angles mentioned above: $60°$ (lower curve) and $90°$.

Note that the minimum value D_m in each case occurs near the $22°$ and $46°$ locations, respectively. As in the case of the rainbow, all possible deviations are present in reality, but it is the "clustering" of deviated rays near the minimum that provides the observed intensity in the halos. (But unlike the case of

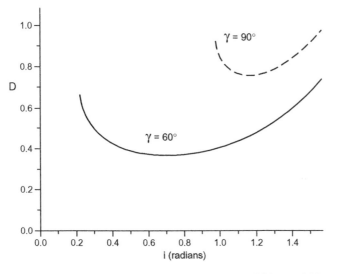

Figure 56.2. $D(i)$ for prism apex angles $\gamma = 60°$ and $90°$

the rainbow, no reflection contributes to their formation in these two cases.) One way to verify analytically that there is a true minimum is, of course, to show from equation (56.3) that when $D'(i) = 0$, $D''(i) > 0$, but this is left as an interesting exercise; it's not as bad as it looks. Note also from the graph that there are restrictions on the angle of incidence i such that outside these, no value of D can be defined for real parameters n and γ. These restrictions arise because of the requirement from equation (56.3) that, in particular,

$$\left\{ n \sin\left[\gamma - \arcsin\left(\frac{\sin i}{n} \right) \right] \right\} \leq 1, \tag{56,4}$$

(i lying within the first quadrant). This inequality places a lower bound on i by unfolding expression (56.3) to obtain

$$i \geq \arcsin\left\{ n \sin\left[\gamma - \arcsin\left(\frac{1}{n} \right) \right] \right\}, \tag{56.5}$$

where it should be noted that $\sin \theta$ is a one-to-one function in the range $(0, \pi/2)$, so that the arcsin function is monotone, increasing in its domain. For $\gamma = 60°$ and $90°$, respectively, this corresponds to $i \geq 13.5°$ and $57.8°$. Regarding the restriction on larger apex angles, it is required from theorem 2 that

$$n \sin\frac{\gamma}{2} \leq 1, \tag{56.6}$$

or

$$\gamma \leq 2 \arcsin\left(\frac{1}{n}\right); \qquad (56.7)$$

for $n = 1.31$ this becomes $\gamma \leq 99.5°$. An apex angle of $120°$ clearly exceeds this; physically this corresponds to total internal reflection within the prism. Arguments of this type enable limits to be placed on the Sun's altitude or the latitude of the observer for certain types of halos to be visible (atmospheric conditions permitting). We illustrate this in question 59 for a beautiful halo known as a *circumzenithal arc*. There are many other halo-type phenomena, some of them very rare. Again, the reader is encouraged to visit Les Cowley's website for more details of these.

There are many more topics we could explore before leaving the subject of ice crystal halos, but the one I choose to end this question concerns why the $46°$ is usually so much fainter than its more commonplace sister, the $22°$ halo. There are several possible contributing mechanisms, the simplest being that the available light scattered through the crystals is directed and distended into a circular image of much larger radius in the former case. An imperfect analogy (but perhaps a helpful one) is that a circular ripple on the surface of a pond or puddle has reduced amplitude as it propagates outward with ever-increasing radius. In this case, of course, the reason is that wave energy is conserved, but likewise, for the halo, the intensity per unit of arc will be less for a given source of light in the case of the larger halo.

Another significant reason is that for the $46°$ halo there is a considerable restriction on the extent of incident rays that can contribute to it, i.e., that can exit the crystal in the direction required. We can find these restrictions on angles of incidence that can give rise to this halo. It is perhaps more helpful visually to work in degrees. From figure 56.3, applying Snell's law of refraction to both the entrance and exit faces, we have that

$$\sin i = n \sin r_1 \text{ and}$$
$$\sin(90° - r_1) = \cos r_1 = n^{-1} \sin r_2. \qquad (56.7)$$

Total internal reflection will occur at the exit face when $r_2 = 90°$, so from the equations above it follows that

$$n \cos r_1 = 1, \qquad (56.8)$$

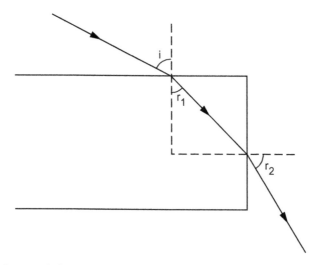

Figure 56.3. Ray path through a 90° prism for the 46° halo

and hence that

$$\sin i = n \sin (\arccos n^{-1}) = \sqrt{n^2 - 1}. \qquad (56.9a)$$

This corresponds to the *minimum* value of i that can give rise to this particular halo, so it follows that

$$i_{min} = \arcsin\left(\sqrt{n^2 - 1}\right). \qquad (56.9b)$$

For a generic value of $n = 1.3$ for ice, $i_{min} \approx 56°$. The maximum value permitted for i is, of course, 90°, i.e., tangential incidence. The implications of this for r_2 are as follows. Clearly, for $i = 90°$

$$r_1 = \arcsin n^{-1}, \text{ so } r_2 = \arcsin (n \cos r_1) = \arcsin(\sqrt{n^2 - 1}). \qquad (56.10)$$

Again, for $n = 1.3$, this implies that the minimum value of r_2 is approximately 56°. Summarizing these results, then, both the angle of incidence *and* the angle of refraction at the exit face lie in the approximate interval $(56°, 90°)$ or, in radian measure, $(0.98, \pi/2)$. Clearly, this is a fairly restrictive domain. The

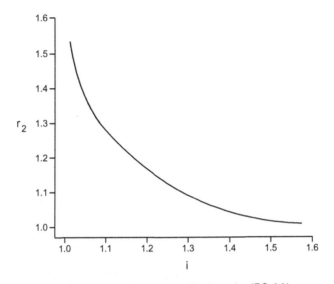

Figure 56.4. The exit angle r_2 (i) given by (56.11)

graph of $r_2(i)$ is given in figure 56.4, using the generalized version of expression (56.10) for r_2, namely

$$\sin r_2 = n \cos r_1 = n \cos \left[\arcsin \left(\frac{\sin i}{n} \right) \right] = \sqrt{n^2 - \sin^2 i}. \qquad (56.11)$$

From figure 56.5 we can also obtain bounds on the relative width of the emergent beam from refraction through a 90° prism. For an arbitrary angle of incidence i (within the permitted interval [51°, 90°]) we know $r_2(i)$ from equation (56.11). If the thickness of the crystal is h, the width W of the of the emergent beam, from figure 56.5, is

$$W = h \cos r_2, \qquad (56.12)$$

so the relative width

$$\tilde{W} = \frac{W}{h} = \cos[\arcsin(n^2 - \sin^2 i)^{1/2}]$$
$$= (1 - n^2 + \sin^2 i)^{1/2}. \qquad (56.13)$$

The obvious requirement that $\tilde{W}$ should be a real number places an even narrower restriction on the domain of i; clearly, for i in the first quadrant this implies that

$$i \geq \arcsin \left| n^2 - 1 \right|^{1/2}. \qquad (56.14)$$

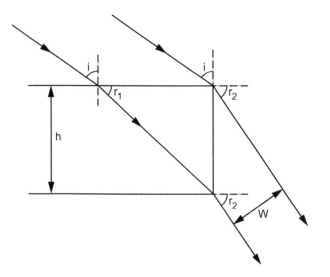

Figure 56.5. Restrictions on the emergent beam width for the 46° halo

For $n = 1.3$ this means that $i \geq 56°$. Thus, for

$$56° \leq i \leq 90°, \; 0 \leq \tilde{W} \leq (2 - n^2)^{1/2} \approx 0.56. \qquad (56.15)$$

Therefore, this restriction on the relative width of the emergent beam, the restriction on incident angles, and the larger circumference of the 46° halo (so the available light is scattered into a larger image) all contribute to the relative faintness of this halo compared with the more common 22° one.

Q.57:   What is that shaft of light above the setting Sun?
[color plate]

It's a *Sun pillar*. Until my university, in its wisdom, built a huge new dormitory behind my office building, I was able to see many wonderful sunsets, and frequently ran outside with my camera to get a permanent record of some of them. Sometimes I saw and photographed some impressive looking Sun pillars. But how are they formed?

As plate-like ice crystals fall they can reflect and refract sunlight. As we have noted already in the previous question and will see in the next question, refraction is the mechanism producing both halos *and* sundogs. However, Sun pillars are produced by *reflection* from the faces of tilted crystals. Specifically, the more common upper pillars result from the reflection of sunlight

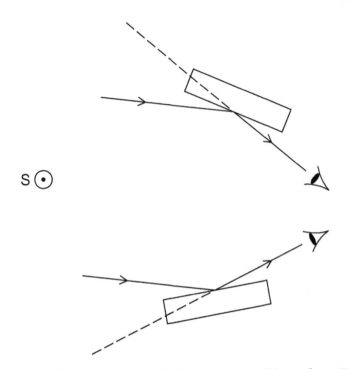

Figure 57.1. Plate crystals contributing to upper and lower Sun pillars

downward from the *lower* faces of such crystals, and lower pillars arise from sunlight being reflected upward from the upper faces of tilted crystals (see figure 57.1). Upper pillars can become brighter as the Sun descends below the horizon, and can appear to extend very high up in the sky if the upper crystals have larger tilts than the lower ones (see figure 57.2(a)). The angular extent of the pillar is about twice the largest tilt angle as can also be inferred from figure 57.2(a). Artificial sources of light can also produce light pillars (figure 57.2(b)).

As can be seen from the geometry of figure 57.3, if the Sun is at an altitude of θ and the lower crystal face is aligned at an angle α to the incoming ray, then the observer will see an image from that ray at an angle $2\alpha + \theta$ to the horizontal. This angle is independent of whether or not the incoming ray is refracted into the crystal, reflected by the inside of the upper surface, and then refracted out to (possibly another) observer, as the figure illustrates. Thus, the reflected ray and the emergent ray are parallel. In principle, such parallel rays could interfere and produce iridescent patterns (like those discussed for oil slicks and butterfly wings), but in practice this is not observed. Why do you think that is?

(a)

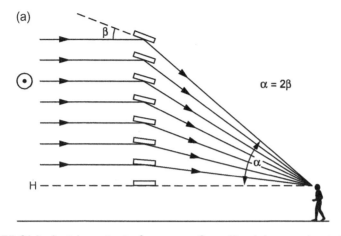

Figure 57.2(a). Angular extent of an upper Sun pillar (observer is at right) (redrawn from Lynch and Livingston (2004))

(b)

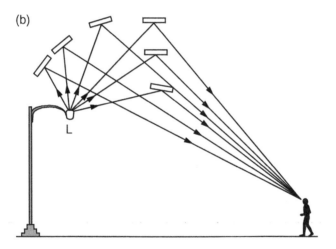

Figure 57.2(b). Light pillars can be formed by nearby light sources (redrawn from Lynch and Livingston (2004))

Q.58: **What is that colored splotch of light beside the Sun?**
[color plate]

Well, it's a *sundog* (or *parhelion* or *mock Sun*), and it is a fairly common sight, but no less fascinating for all that. The basic mechanism of formation is shown in figures 56.1(b) and 58.1. Sundogs are formed by the refraction of light through downward-drifting horizontally aligned hexagonal plate crystals, and

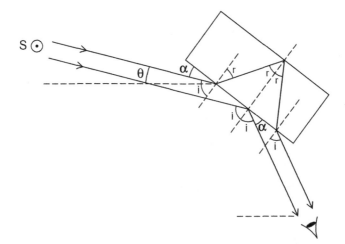

Figure 57.3. Ray paths contributing to an upper Sun pillar

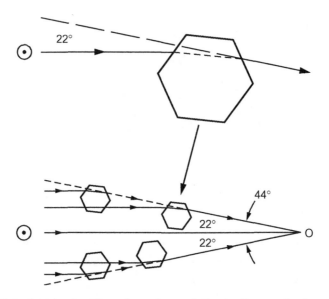

Figure 58.1. Angular location of sundogs relative to the sun (redrawn from Lynch and Livingston (2004))

can on occasion be extremely bright. They are frequently observable when the Sun is setting, close to the position of the 22° halo (see question 56; the mathematics involved is identical), but usually just outside it (if the halo is visible). Always seen at the same altitude as the Sun, at higher solar altitudes they can be considerably far outside the halo location. This is because the rays forming them become increasingly skewed relative to the crystal axis, and the ray paths of minimum deviation are no longer traversed inside the crystal, also resulting sometimes in long "tails" pointing away from the Sun. Depending on the position of the cirrostratus clouds relative to the Sun, sundogs may be seen on either side of the Sun at the same time, or on just one side.

What different formation mechanism gives rise to the sundog as opposed to many of the types of ice crystal halos that may occur, specifically the common 22° halo, since both are caused by approximately hexagonal crystals? The answer is that for halo formation to occur, there must be a fairly uniform distribution of crystal orientations, a result of tumbling *columnar* crystals, refracting light in every possible direction, whereas (as noted above) sundogs are formed by larger plate-like crystals falling with their refracting sides approximately vertical. The reason for the difference in "stability" as it were, is that the smaller columnar crystals are more prone to tumbling; larger ones fall much like a sky-diver does: steadily and with fixed orientation.

Q.59:   What's that "smiley face" in the sky?

I was heading back to my office with a cup of coffee gleaned from the Department of Civil and Environmental Engineering across campus. It was a beautiful crisp day in late November, with thin cirrus clouds scattered about an otherwise clear blue sky. I had nearly reached my office when I heard my name being called. "Professor Adam—look at the rainbow high in the sky!" A student in my Mathematics in Nature class directed my attention to this beautiful, high-altitude arc. From the discussion of rainbow formation in the class, she *should* have known it was not a rainbow, because (i) there was no rain, (ii) it was far too high in the sky to be a rainbow, and finally, (iii) it was like an upward smiley face, not a downward one! It was, in fact, a *circumzenithal arc* (CZA). On the other hand, she had noticed it, and I had not, being too busy looking at the sundogs—yes, both of them! I rushed into my office to retrieve my camera, and "my" CZA appeared some time later on the Earth Science Picture of the Day web site (*http://epod.usra.edu/*), along with several other pictures taken during the same display. My description was as follows:

> This circumzenithal arc (CZA) was part of a magnificent display of arcs and parhelia observed during the late afternoon of November 27th, in Norfolk, Virginia. The CZA and parhelia were photographed within minutes of each other. A faint upper tangential arc was also visible for a short while. Later

that afternoon, just after sunset, a faint "sundog pillar" appeared to the north of the setting point; according to Dr. Les Cowley, this strange sight may have been the result of an overlapping parhelion and subparhelion, giving rise to the appearance of a parhelion pillar (another possibility is a variety of Lowitz arc).

The CZA forms only when the Sun is less than 32.3° high, and its appearance changes as the Sun sets: near the upper altitude limit the arc is faint and diffuse and almost in a zenithal position, but as the Sun sets the arc moves away from the zenith (remaining symmetrically convex toward the Sun), and it becomes narrower, while its angular extent increases somewhat. I think that the Sun's altitude was between 25 and 30 degrees at the time this photograph was taken. The cirrus cloud crystals from which this CZA resulted dispersed shortly thereafter and the arc disappeared from sight. CZAs form when sunlight passes through the uppermost basal faces of oriented, hexagonal plate crystals and then exits through vertical side faces. These two faces are perpendicular, causing the colors to disperse more widely than for smaller "prism angles." Consequently, there is little overlap between the colors, and they are very pure hues, purer than those of a rainbow.

The circumzenithal arc is a circular halo that is centered on the zenith point of the observer. There is a nice description of the phenomenon on Les Cowley's website (see references): "The circumzenithal arc is the most beautiful of all the halos. The first sighting is always a surprise, an ethereal rainbow fled from its watery origins and wrapped improbably about the zenith. It is often described as an 'upside-down rainbow' by first timers. Someone also charmingly likened it to 'a grin in the sky.' Look straight up near to the zenith when the sun is fairly low and especially if sundogs are visible. The centre of the bow always sunwards and red is on the outside."

Now for a little geometry and trigonometry. The sunlight enters from the top of the ice crystal (as shown in figure 59.1) and exits from the side.

We can derive a condition on the Sun's altitude θ for the halo to be visible, i.e., θ must be greater than (or less than) a certain angle, and will assume, as before, that the refractive index of the ice is 1.31. From figure 59.1 we can see that no ray will enter the observer's eye if total internal reflection were to take place at the exit point. This places restrictions on the angle of internal incidence (measured from the normal to the vertical face) θ_2 and hence on the external angle of incidence to the horizontal face, θ_1. Total internal reflection will occur if $r_2 \geq \pi/2$, so taking the equality as the limiting case we have

$$\sin \theta_2 = n_2 \sin r_2 = n_2 = n_1^{-1} = (1.31)^{-1} \approx 0.76, \qquad (59.1)$$

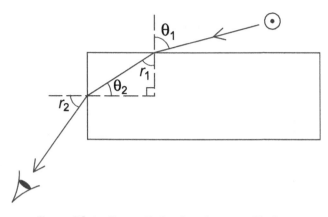

Figure 59.1. Ray path for the circumzenithal arc

since this is an ice-to-air interface. Then

$$\theta_2 \geq \arcsin{(0.76)} \approx 49.7°. \tag{59.2}$$

Therefore, $r_1 \leq 40.3°$, and for the air–ice interface, $\sin{\theta_1} = n_1 \sin{r_1}$, so total internal reflection occurs when

$$\theta_1 \leq \arcsin{[1.31 \sin{(40.3°)}]} \approx 57.9°. \tag{59.3}$$

Under these circumstances no ray enters the observer's eye, and therefore the circumzenithal arc can be seen been only when $\theta_1 \geq 57.9°$, i.e., when the Sun's altitude $\theta \ (= 90° - \theta_1)$ is less than about 32°. Similar kinds of calculation apply to other halo types to yield the corresponding limitations on solar elevation (and hence latitudes of visibility) or perhaps crystal orientation. In fact, calculations for the *circumhorizontal arc* are complementary to those presented here (see Adam (2006) for details). This arc, as its name implies, is parallel to the horizon; again, it is colored because of dispersion, and can be one of the most spectacular arcs there is to witness (the reader may have encountered one such display via email; it is misleadingly labeled a "fire rainbow", and I receive a picture of it about once a month!). It is formed as a result of light being refracted through the vertical face of many horizontal hexagonal ice crystal plates, and then out through the lower base. It can only occur for solar elevations greater than 58° (the complement of the 32° angle discussed above), and this of course imposes limitations on the latitudes in which this particular arc can be seen.

Q.60:   **What are those colored rings around the shadow of my plane?** [color plate]

It's called a *glory*. And before you object, I do realize that flying is not the same as walking, but my excuse is that the same phenomenon can be seen with our feet firmly planted on the ground! Mountaineers and hill climbers have noticed on occasion that when they stand with their backs to the low-lying Sun and look into a thick mist below them, they may sometimes see a set of colored circular rings (or arcs thereof) surrounding the shadow of their heads. Although an individual may see the shadow of a companion, the observer will see the rings only around his or her head. Again, many details of glories can be found from Les Cowley's website (*http://www.atoptics.co.uk/*).

During the nineteenth century, many such observations of the glory were made from the top of the Brocken mountain in central Germany, and it became known as the brocken bow or the "Specter of the Brocken" (being frequently observed on this high peak in the Harz mountains of central Germany). It also became a favorite image among the Romantic writers; it was celebrated by Coleridge in his poem "Constancy to an Ideal Object." Other sightings were made from balloons, the glory appearing around the balloon's shadow on the clouds. Nowadays, while not noted as frequently as the rainbow, it may be seen most commonly from the air, with the glory surrounding the shadow of the airplane. Once an observer has seen the glory, if looked for, it is readily found on many subsequent flights (provided one is on the shadow side of the aircraft!).

This phenomenon can be understood in the simplest terms as essentially the result of light backscattered by cloud droplets, the light undergoing some unusual tranformations en route to the observer (with a correspondingly complicated mathematical description), transformations that are *not* predictable by standard geometrical optics, unlike the basic description of the rainbow. That this must be the case is easily demonstrated by noting a fallacy present in at least one popular meteorological text (e.g., Ahrens' *Meteorology Today*). The glory, it is claimed, is formed as a result of a ray of light tangentially incident on a spherical raindrop being refracted into the drop, reflected from the back surface and reemerging from the drop in an exactly antiparallel direction into the eye of the observer (see figure 60.1).

If such a picture is correct, then since the angle of incidence of the ray is 90°, it follows by Snell's law of refraction that the angle of refraction is

$$r = \arcsin\left(\frac{1}{n}\right), \tag{60.1}$$

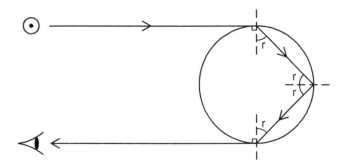

Figure 60.1. An incorrect ray path for the glory

where n is the refractive index of the raindrop. For an air/water boundary, $n \approx 4/3$ and so $r \approx 48.6°$ (ignoring the effects of dispersion here, though this does occur as noted above; note also that for a water/air boundary the reciprocal of n must be used). This means (by the law of reflection) that at the back of the drop, the ray is deviated by more than a right angle, since $2r \approx 97.2°$, and by symmetry the angle of incidence *within* the drop for the exiting ray is also 48.6°, so the total deviation angle (as we saw for the primary rainbow) is

$$D(i) = \pi + 2i - 4\arcsin\left(\frac{\sin i}{n}\right)$$

$$= -4\arcsin\left(\frac{3}{4}\right) \approx -194.4° \text{ or } +165.6°, \qquad (60.2)$$

(modulo 2π) since $i = 90°$. And this means that the exiting ray is about 14° short of being "antiparallel." It just won't work as a mechanism for the glory! There are basically two potential ways out of this. We could ask what value of refractive index n would be necessary for the diagram to be correct; thus, $i = 90°$ as before but now $r = 45°$; this means that

$$n = \frac{\sin i}{\sin r} = \sqrt{2} \approx 1.4. \qquad (60.3)$$

Something between water and glass might do it, perhaps plastic! But absent any evidence that clouds are composed of transparent plastic spheres we discard this suggestion. What remains? Another possibility is that somehow the ray travels around the surface (as a surface wave) for part (or parts) of its trip, the surface portion comprising the missing piece θ,

$$\theta = 180° - 2(180° - 2r) = 4r - 180° \approx 14.4° \qquad (60.4)$$

for $r \approx 48.6°$. The resulting path in the droplet need not be symmetric to account for an antiparallel exiting ray. A detailed mathematical study of the glory is, regrettably, far too complicated to go into here, but the interested reader is referred to the references for further details. In addition, there is an excellent site by Philip Laven in which he discusses many optical phenomena, but particularly rainbows and glories, and he has developed a very convincing explanation for the optical glory. His website can be found in the references.

Exercise: Derive equation (60.4).

Q.61:   Why does geometrical optics imply infinite intensity at the rainbow angle?

To end this "in the sky" section, we consider the concern raised in question 49, namely that geometrical optics actually breaks down at the rainbow angle, the deviation angle corresponding to the direction of the rainbow (of any particular order).

Consider a thin "bundle" of rays impinging a spherical raindrop (of radius a) as shown in figure 61.1(a) and its inset. The angles of incidence of the rays comprising this bundle lie in the interval $(i, i + \delta i)$ The diagram is cylindrically symmetric about an axis parallel to the direction of the incoming rays, so the area "seen" by the cylindrical ray bundle is

$$\delta A \approx (2\pi a \sin i)(a \cos i)|\delta i| = \pi a^2 |\delta i| \sin 2i. \qquad (61.1)$$

These rays are scattered into an angular interval $(\theta, \theta + \delta\theta)$, which occupies a solid angle to be determined below. Associated with this small interval is an "onion ring" surface area element δS, where from figure 61.1(c)

$$\delta S \approx (2\pi a \sin \theta)(a|\delta\theta|) = 2\pi a^2 \sin \theta |\delta\theta|, \qquad (61.2)$$

which is equivalent to a solid angle element

$$\delta\Omega = 2\pi \sin \theta |\delta\theta|. \qquad (61.3)$$

Now if I_0 is the rate at which the incident light energy falls on a unit area perpendicular to its incoming direction, then the corresponding rate at which it enters a unit solid angle on emerging from the drop is

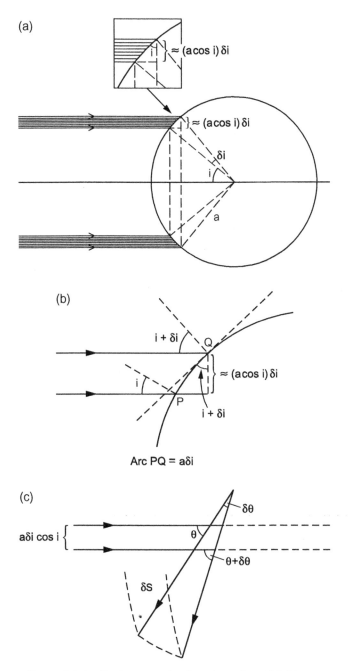

Figure 61.1. Basic geometry for scattering of light rays

$$I \approx I_0 \frac{(\pi a^2 \sin 2i)|\delta i|}{(2\pi \sin \theta)|\delta \theta|},$$

(61.4)

and passing to the limit, we find that the relative energy rate is

$$\frac{I}{I_0} = \frac{a^2 \sin 2i}{2 \sin \theta} \left| \frac{d\theta}{di} \right|^{-1}.$$

(61.5)

In the rainbow problem discussed in question 49, the deviation of an incoming ray is denoted by $D(i)$, and that is just $\pi - \theta(i)$ here; so the condition for an extremum in $D(i)$ (defining the rainbow angle) is just $D'(i) = 0$, which immediately implies that $\theta'(i) = 0$, and explains the singularity arising at the rainbow caustic when geometrical optics is used.

There is still a little more to be said as far as geometrical optics is concerned, based on equation (61.5). It would appear that equation (61.5) also predicts an infinite intensity when, in degrees, $\theta = 0°$ and $180°$ (i.e., $D(i) = 180°$, and $0°$, respectively). But θ (and hence D) takes on these values only when the angle of incidence $i = 0$, and, specifically, for one internal reflection (for which $\theta = 0°$)

$$\frac{\sin 2i}{\sin \theta} \approx \frac{2i}{\theta}.$$

(61.6)

But when $i \approx 0°$

$$\theta = 180° - D(i) \approx 180° - (2i - 4r + 180°) = 4r - 2i.$$

(61.7)

Therefore, since now Snell's law reduces to $i \approx nr$,

$$\frac{\sin 2i}{\sin \theta} \approx \frac{i}{2r - i} \approx \frac{nr}{2r - nr} = \frac{n}{2 - n}.$$

(61.8)

For water, $n \approx 4/3$ and this ratio is then equal to two. From equation (61.7)

$$\frac{d\theta}{di} = 4 \frac{dr}{di} - 2 = \frac{4}{n} - 2 = 1$$

(61.9)

in this limit of small angles of incidence, so there is no significant increase in I in this situation.

Finally, it is appropriate to note that a better approximation to the actual intensity distribution in the vicinity of the rainbow angle is provided by so-called Airy theory. This, though not a perfect observational fit, incorporates the wave nature of light, and was developed and in 1838 published by the then Astronomer Royal in Britain, Sir George Biddell Airy. Unfortunately, it would take us too far afield to discuss this fascinating improvement to the mathematical model of the rainbow.

In the nest

Q.62: How can you model the shape of birds' eggs?

My house has a front porch with three pillars, and the inside-facing part of each pillar, at the top, is a popular spot for birds to build nests; so far we've been host to doves, sparrows, and finches. However, a strong gust of wind in the right direction often demolishes the nests, sadly. But the parent birds are learning: they have decided that the hanging flower baskets are more desirable locations from which to raise a family, and, at the same time, frustrate the local cat population. Great care has to be taken when getting the "host" basket down to water the flowers: several eggs can usually be found in the nest. It's just a question of location, location, location . . .

I was interested to learn that the geometric properties of birds' eggs are sometimes useful for computations of shell permeability and incubation times, and even the design of egg trays for hens' eggs. It should not be surprising, therefore, that there are mathematical models of the shape of birds' eggs and the relationship between their surface area and volume. For any closed surface, there is a relationship between its area A and volume V of the following form:

$$A = kV^{2/3}, \tag{62.1}$$

k being a dimensionless constant depending on the shape of the closed surface. This is obvious from dimensional considerations; both sides must have dimensions of (length)^2, since volume V and surface area A scale respectively as the cube and the square of a linear dimension. It is easy to see that for a cube, $k = 6$. This is related to a very useful quantity called the sphericity index, because it is a measure of how close to spherical is a particular three-dimensional shape; see question 63 for further details. In questions 64–67 we shall examine mathematical models dominated by, respectively, trigonometric, algebraic, calculus, and geometric concepts; who knew that there are so many ways to look at an egg?

Question 62. House finch eggs; the nest is in a hanging flower basket.

Exercise: Show that for a sphere, $k = \sqrt[3]{36\pi} \approx 4.836$. Then read question 63 below!

Q.63:   What is the sphericity index?

A useful measure of volume relative to surface area is called the *sphericity index*. Since volume V and surface area A scale respectively as the cube and the square of a linear dimension, it is evident that a suitable *dimensionless* measure of the volume-to-surface area ratio for an object should involve some power of the ratio $V^{2/3}A^{-1}$. The sphericity index is one such measure, being defined as

$$\chi = \frac{4.836\,V^{2/3}}{A}. \tag{63.1}$$

Hmm . . . What's with the 4.836? After all, it's hardly a household number. This arises by requiring that the sphericity index for a sphere is unity, so if we temporarily replace the above number by α, it follows that for a sphere of radius R

$$\chi = 1 = \frac{\alpha V^{2/3}}{A} = \alpha\frac{\left(\frac{4}{3}\pi R^3\right)^{2/3}}{4\pi R^2} = \alpha\frac{\left(\frac{4}{3}\pi\right)^{2/3}}{4\pi}, \tag{63.2}$$

from which we obtain $\alpha = (4\pi)^{1/3} \cdot 3^{2/3} \approx 4.836$. Because a sphere has the largest volume-to-surface area ratio for any closed surface it follows that for other shapes, $0 < \chi < 1$. Let's consider some examples: for a cube it is readily shown that $\chi \approx 0.806$; for two "kissing" spheres (i.e., in tangential contact) $\chi = 2^{-1/3} \approx 0.794$. I like the sphericity index because it is always of order one; there is another quantity that is sometimes used in biological contexts, called the *flatness index* γ, where

$$\gamma = \frac{A^3}{V^2} \propto \chi^{-3}, \tag{63.3}$$

and it is generally for that reason larger than χ, being approximately 113 for a sphere and 216 for a cube. The observant reader will have noticed immediately for all these examples that (1) the radius of the sphere of side length of the cube has not been specified; and (2) it is not necessary to do so, because all the dimensional quantities cancel out (by design).

Q.64: Can the shape of an egg be modeled trigonometrically?

Regarding the shape of eggs, one investigation involved a set of measurements of A and V for eggs of different shapes and sizes, and found empirically that for most eggs $k \approx 4.951$; i.e., marginally more "cuboid" than a sphere! And while it is clear that eggs rarely possess the symmetry of prolate spheroids, nevertheless the latter can be a good approximation. To that end, we will determine k for just such an object: a prolate spheroid being an ellipsoid of revolution generated by rotating an ellipse around its *major* axis (rotation about the minor axis produces an *oblate* spheroid). The surface area of such ellipsoids is readily calculated (see equation (64.4)). Once k is determined theoretically, the accuracy of the ellipsoid model can be estimated by measuring the volume of the egg by indirect methods such as total immersion. If the error is a matter of only a few percent, then the surface area can be computed with confidence to a similar degree of accuracy.

Consider then an ellipse with semimajor axis length a and semiminor axis length $b < a$ (figure 64.1) so the equation of the ellipse in the (x–z) plane is

$$\frac{x^2}{b^2} + \frac{z^2}{a^2} = 1. \tag{64.1}$$

The corresponding prolate ellipsoid has the equation

$$\frac{x^2 + y^2}{b^2} + \frac{z^2}{a^2} = 1. \tag{64.2}$$

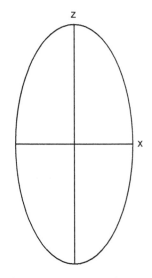

Figure 64.1. Ellipse with axes of length a and b < a

The parameter describing the "elongation" in the literature is $a/b = p^{-1}$; clearly, the eccentricity ϵ of the ellipsoid is related to p via the relation $\epsilon = \sqrt{1 - p^2}$. Note that ϵ is defined as c/a, where the foci of the ellipse (oriented vertically) are at $(0, \pm c)$, and $c^2 = a^2 - b^2$. Now from elementary calculus the volume of the prolate spheroid is

$$V = \frac{4}{3}\pi a b^2 = \frac{4}{3}\pi p^2 a^3, \tag{64.3}$$

and its surface area is

$$A = 2\pi b\left(b + \frac{a}{\epsilon}\arcsin \epsilon\right) = 2\pi a^2\left(p^2 + \frac{p}{\sqrt{1-p^2}}\arcsin \sqrt{1-p^2}\right). \tag{64.4}$$

From equation (62.1), k is therefore defined in terms of p alone as

$$k = Cp^{-1/3}\left(p + \frac{\arcsin \sqrt{1-p^2}}{\sqrt{1-p^2}}\right), \tag{64.5}$$

C being the constant $2\pi(4\pi/3)^{-2/3} \approx 2.418$. The graph of $k(p)$ is shown in figure 64.2. It is of interest to consider k in the limit of $p \to 1^-$; this is the limiting case of a sphere. Although $k(1)$ is indeterminate, the limit does exist; it is $k = 2C \approx 4.836$, a number encountered above.

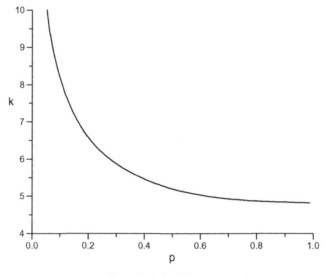

Figure 64.2. $k(p)$, where $p = b/a$

Exercise: Show that equation (64.5) is equivalent to

$$k = \left(\frac{9\pi}{2p}\right)^{1/3}\left(p + \frac{\arccos p}{\sqrt{1-p^2}}\right). \tag{64.6}$$

If we also imagine a sphere of radius a circumscribing the spheroid, then the former has surface area $S = 4\pi a^2$, and the ratio of the areas is

$$\frac{A}{S} = \frac{1}{2}\left(p^2 + \frac{p}{\sqrt{1-p^2}}\arcsin\sqrt{1-p^2}\right). \tag{64.7}$$

This is also plotted in figure 64.3; in the range shown it is almost a linear function, and clearly approaches unity as $b \to a$ and the sphere is recovered.

No avian eggs are spherical, to my knowledge (this seems to be a consequence of the egg-laying procedure). The elongations a/b range from about 1.19 to about 1.64, corresponding to p in the range 0.84 down to 0.61, there being, it seems, a pronounced concentration between 0.70 and 0.75. As may be seen from the graph of $k(p)$ in figure 64.2, in this range k varies quite slowly: thus $k(0.70) \approx 4.94$ and $k(0.75) \approx 4.90$. This range is not far above the value for a sphere, 4.836, noted above. Some hummingbird eggs depart considerably from the prolate spheroidal form, being shaped more like beans,

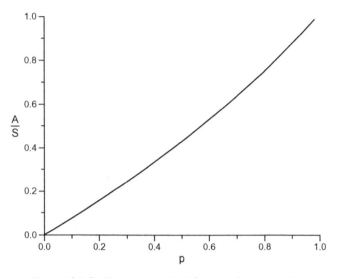

Figure 64.3. The area ratio A/S as a function of p

and one model of the egg of the hummingbird *Stellula calliope* consists of two hemispheres of radius r capping a cylinder of length l and diameter d, though the geometry is reminiscent of a "gelcap" tablet in pain relief products found in drugstore aisles!

Q.65:   Can the shape of an egg be modeled algebraically?

On a purely mathematical level, it is interesting to see how coordinate transformations (a more precise word than distortions!) on a circle can produce interesting egg-like cross-sectional shapes. Consider the equation of the unit circle: $X^2 + Y^2 = 1$. Now we proceed to make coordinate transformations of the (X, Y)-plane according to the rather general requirement that

$$X = \frac{x}{a}, \; Y = \frac{y}{af(X)}, \tag{65.1}$$

so the unit circle becomes, in the new coordinates

$$x^2 + \frac{y^2}{[f(x/a)]^2} = a^2, \tag{65.2}$$

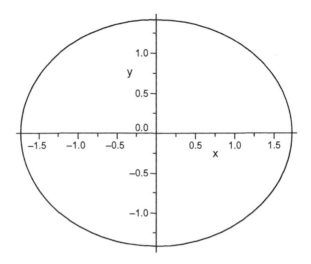

Figure 65.1. Unit circle after transformation $f = k = b/a$

or, restricting ourselves to $y \geq 0$,

$$y = af\left(\frac{x}{a}\right)\sqrt{1 - \frac{x^2}{a^2}}.$$

(65.3)

Since the eggs are axisymmetric (about the x-axis), there is no loss of generality in doing this. Two simple transformations are as follows: (i) $f(X) = k = b/a$ yields the ellipse

$$\frac{x^2}{a^2} + \frac{y^2}{b^2} = 1,$$

(65.4)

and (ii) $f(X) = k + cX$ gives the "ovoid"

$$x^2 + \frac{y^2}{\left(k + c\frac{x}{a}\right)^2} = a^2,$$

(65.5)

or

$$y = a\left(k + c\frac{x}{a}\right)\sqrt{1 - \frac{x^2}{a^2}}.$$

(65.6)

Figure 65.1 shows the parameter choices $a = \sqrt{3}$ and $b = \sqrt{2}$, and figure 65.2 is for $c = 1/\sqrt{3}$ and $k = \sqrt{2/3}$ with the same value of a. The straight lines are images of the lines $Y = \pm$ constant under this coordinate transformation.

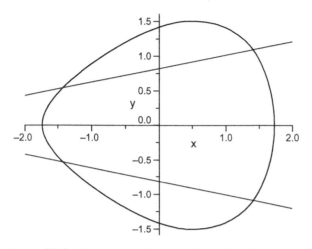

Figure 65.2. Unit circle after transformation $f = k + cX$

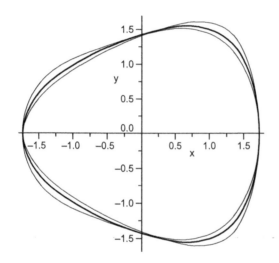

Figure 65.3. Unit circle after transformation $f = k + cX + dX^2$

Figure 65.3 is based on the quadratic choice for $f(x)$ for various values of d (and the same choices for a and c):

$$y = a\left(k + c\frac{x}{a} + d\frac{x^2}{a^2}\right)\sqrt{1 - \frac{x^2}{a^2}}, \text{ i.e.,} \tag{65.7}$$

$$x^2 + \frac{y^2}{\left(k + c\frac{x}{a} + d\frac{x^2}{a^2}\right)^2} = a^2. \tag{65.8}$$

Q.66:   Can the shape of an egg be modeled using calculus?

Another very interesting technique for distorting an ellipse into an ovoid egg shape was developed by Carter in studying the hen's egg (see the references). The degree of skewness of the egg is built in as one of the parameters of the model. The initial ellipse is to be modified to represent a cross section through the axis of symmetry of the egg. Obviously, such a cross section is not elliptical, as evidenced by the fact that the maximum width of the egg (perpendicular to the axis) does not occur at the midpoint of the axis. The transformation applied elongates the ellipse toward one pole and compresses it toward the other.

For an ellipse centered at the origin with major axis scaled to unit length ($a=1$) and minor axis of length $b<1$, the y-coordinate of the upper ($+$) and lower ($-$) halves is given by

$$y = \pm b\left(\tfrac{1}{4}-x^2\right)^{1/2}. \tag{66.1}$$

The origin can be shifted to the "sharper" end (say) by the substitution $X = x+1/2$, whereupon

$$y = \pm b\left[\tfrac{1}{4}-\left(X-\tfrac{1}{2}\right)^2\right]^{1/2}, \tag{66.2}$$

where $X \in [0, 1/2]$. Asymmetry, or skewness, can be introduced by the non-linear variable change $z=X^p$, $p>0$. Note that $z(0)=0$ and $z(1)=1$, so the major axis length is preserved under this transformation. Furthermore, $z > X$ if $0 < p < 1$ and $z < X$ if $p > 1$. Then, when X is formally replaced by z, the expression (66.2) becomes, in terms of X,

$$y = \pm b\left[\tfrac{1}{4}-\left(X^p-\tfrac{1}{2}\right)^2\right]^{1/2}. \tag{66.3}$$

(This expression is plotted in figure 66.1.) Clearly, the extrema of y occur at $X=X_m$, where $X_m^p = 1/2$, i.e., when

$$p = \frac{\log 2}{\log\left(1/X_m\right)}, \tag{66.4}$$

allowing an estimate of p from measurements of the egg.

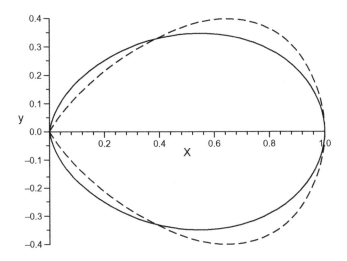

Figure 66.1. Graph of equation (66.3) for two values of b and p

Regarding the egg as a solid of revolution, with axis of symmetry X, the egg volume is

$$V = \pi \int_0^1 y^2(X)\,dX$$

$$= \pi b^2 \int_0^1 \left(\frac{1}{4} - \left(X^p - \frac{1}{2} \right)^2 \right) dX$$

$$= \pi b^2 \int_0^1 (X^p - X^{2p})\,dX$$

$$= \frac{\pi b^2 p}{(p+1)(2p+1)}. \tag{66.5}$$

When $p=1$ this reduces to $V = \pi b^2/6$, the volume of a prolate spheroid with unit major axis and minor axis of length b. The surface area S of this ovoid may also be determined, though not generally in closed form:

$$S = 2\pi \int_0^1 y\sqrt{1 + \left(\frac{dy}{dX} \right)^2}\,dX$$

$$= \pi b \int_0^1 \left[1 + 4\left(X^p - \frac{1}{2} \right)^2 (b^2 p^2 X^{2p-2} - 1) \right]^{1/2} dX. \tag{66.6}$$

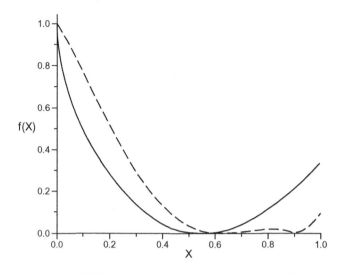

Figure 66.2. Graph of $f(X)$ for $p = 1.16$, $b = 0.7$ and $p = 1.5$, $b = 0.7$ (dashed curve)

While this integral may be generally difficult to evaluate in closed form, some progress can be made if the second term is less than one in magnitude. To pursue this, we abbreviate the expression for S by writing it as

$$S = \pi b \int_0^1 [1 + f(X)]^{1/2} dX, \qquad (66.7)$$

where $f(X)$ is obviously defined from the integrand above. If $|f(X)| < 1$, then the integrand can be expanded in a convergent power series, and integrated termwise. Depending on the magnitude of $|f(X)|$, it may be necessary to retain only a few terms of the expansion.

It can be seen from figure 66.2 that for $X \in (0, 1)$, $|f(X)| < 1$, and so for most of the range of X the series for the integrand will converge quite rapidly. Thus,

$$S = \pi b \int_0^1 \left(1 + \frac{1}{2}f - \frac{1}{8}f^2 + \frac{1}{16}f^3 - \frac{5}{128}f^4 + \cdots \right) dX, \qquad (66.8)$$

each term of which is directly integrable, insofar as it involves powers of X.

Exercise: Investigate properties of $|f(X)|$ for general values of b and p.

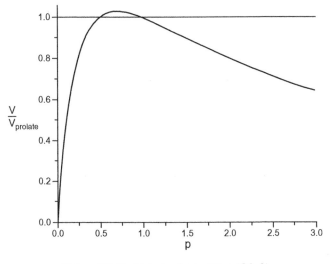

Figure 66.3. Graph of equation (66.9)

Exercise: Show that, if the ovoid's long axis were not scaled to unity, but were instead of length L, the corresponding values of p, V, and S now become $\tilde{p}$, $\tilde{V}$, and $\tilde{S}$, where

$$\tilde{p} = \frac{\log 2}{\log (L/X_{\mathrm{m}})}; \tilde{V} = L^3 V \text{ and } \tilde{S} = L^2 S.$$

Carter defines two important parameters in connection with the shape of eggs: the *shape index* and the *skewness index*. The former is defined as the ratio of the maximum breadth of the ovoid to its length. The latter is the distance from the flatter end of the ovoid to its line of maximum width, divided by its length. The reason for this is an eminently practical one: when using callipers to measure the egg, it is easier to start from the flatter end!

Let's now compare the relative volumes of the model hen's egg as a function of p. When $p=1$ the cross section is elliptical, and the egg is a prolate spheroid with volume $V = \pi b^2/6$. There is a narrow range of p where the ratio

$$\frac{V}{V_{\mathrm{prolate}}} = \frac{6p}{(p+1)(2p+1)} \tag{66.9}$$

exceeds one, as can be seen from figure 66.3. This is easiest to see if we examine the difference

$$D = \frac{V - V_{\mathrm{prolate}}}{\pi b^2} = \frac{p}{(p+1)(2p+1)} - \frac{1}{6}. \tag{66.10}$$

Clearly, $D > 0$ when

$$2p^2 - 3p + 1 = (2p - 1)(p - 1) < 0, \tag{66.11}$$

i.e., when $1/2 < p < 1$; for $0 < p < 1/2$ or $p > 1$, $D < 0$ and the volume of the prolate spheroid exceeds that of the ovoid.

Carter found that measurement of the volume of hens' eggs by water displacement showed that almost invariably they are smaller by volume than the corresponding prolate spheroid, drawing the conclusion that $p > 1$; however, it is clear from the above algebra that $0 < p < 1/2$ is also a possibility, though in practice this just means that the origin of the X-axis is located at the flatter end of the egg.

Q.67:   Can the shape of an egg be modeled geometrically?

A recent paper by D. E. Baker (see the references) on the topic of the shape of bird eggs uses an approach using projective geometry, and is based on so-called *path curves*. The mathematical concept underlying path curves is part of projective geometry; space (whether two or three dimensional) is "mapped" or "projected" onto itself. A path curve is the locus of a point that is repeatedly moved by a linear transformation. It was developed from work of Rudolph Steiner by Lawrence Edwards, who was interested in generating shapes prevalent in nature: buds, eggs, ventricles, and even embryo shapes. A detailed discussion of this would take us too far afield; suffice it to say that the paper is based on path curves in two-dimensional space; further details and citations may be found in that article.

The essential idea is that the oval profile of the egg cross section is determined by geometric sequences of points (and the lines joining them) on two lines perpendicular to the oval axis of symmetry. It is claimed that the resulting equation is the simplest analytic description currently available for a bird egg profile. To derive this equation, consider figure 67.1 with the egg axis of symmetry located in the Cartesian plane on the x-axis interval $[-1, 1]$. An arbitrary point on the egg profile is defined by its coordinates (x, y) and using similar triangles in the figure, it can be seen that the y-coordinates of the points A' and A are given by

$$A_1 = \frac{2y}{1+x} \text{ and } A_2 = \frac{2y}{1-x}. \tag{67.1}$$

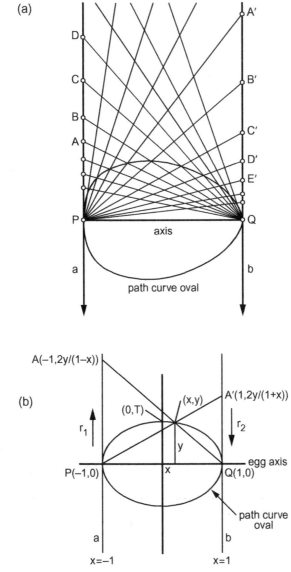

Figure 67.1. Path curve oval and geometry for A and A' (redrawn from Baker (2002))

We use a parameter t that, as indicated in figure 67.1, steps out the points of intersection of the lines PA' and QA; these points lie on the oval profile. Thus, let

$$\frac{2y}{1-x} = cr_1^t \text{ and } \frac{2y}{1+x} = dr_2^{-t}; \tag{67.2}$$

these are the parametric equations for the sequences of points on the vertical lines at P and Q, respectively, i.e., for the points A and A', respectively. When $t = 0$ the initial locations are at c and d, respectively. From these two equations the parameter t may be eliminated to obtain the equation

$$\lambda \ln\left(\frac{2y}{c(1-x)}\right) + \ln\left(\frac{2y}{d(1+x)}\right) = 0, \tag{67.3}$$

where $\lambda = (\ln r_2)/(\ln r_1)$. Upon inverting this expression to solve explicitly for y we find

$$y = T(1-x)^{\lambda/(1+\lambda)}(1+x)^{1/(1+\lambda)}, \tag{67.4}$$

where $T = \frac{1}{2}(cd)^{1/(1+\lambda)}$. The profile of the egg is determined by this path curve; different shapes are determined by the λ and T shape parameters. The former is a measure of the egg asymmetry, and T is a scaled equatorial radius of the egg (perpendicular to the axis of symmetry). Both are clearly related to the natural geometric shape of an egg. In particular, λ is more natural and useful for comparison of egg shapes with models than some of the constants arising in other models. The precise numerical details need not concern us here; suffice it to point out that for $\lambda = 1$ the path curve oval is an ellipse, and for $\lambda > 1$ one end of the path curve is blunter and one is sharper. Baker used a statistical method (least-squares nonlinear regression) with respect to these two parameters to find those path curves that best fit some 250 species of egg with considerable accuracy.

Closing comment

The last four questions have dealt with different approaches to modeling the shapes of birds eggs. It might be of interest to compare these models to one another and to existing data about real eggs. Depending on the amount of effort devoted to it, a project such as this could evolve into something quite substantial; it is left therefore as "an exercise for the interested reader"!

In (or on) the water

Q.68: What causes a glitter path? [color plate]

Have you ever noticed golden "cylinders of Sun" (a "glitter path") on the water as the Sun sets in the west or rises in the east? This beautiful "liquid gold" effect arises, of course, from the reflection of sunlight from the surface of the water, but it is not quite *that* simple. This phenomenon is not reflection from a flat surface, in general—the surface usually is rippled with waves. And since each of these ripples can be a reflecting surface for sunlight (or other light sources), a glitter path is composed of myriads of tiny "glints" or distorted images of the Sun. In fact, the mechanism producing a glitter path is very similar to that producing a light pillar (see question 57). Of course, there are often local variations in the "roughness" of the surface: a body of water may be relatively smooth in part, with other regions roughened in the form of "cat's paws" by wind gusts, and yet other parts may have slicks limiting the formation of ripples. Therefore, any given observation, like much else in nature, is certainly unique. And being a reflection phenomenon, the glitter colors near sunset can be especially beautiful, perhaps gold, or crimson, or even salmon pink.

We'll note the basic common features of glitter paths and then use some simple geometric arguments to explain them. Seen from the air, the glitter path is generally elliptical. An observer at the edge of a body of water will therefore usually see an elongated ellipse; this is the glitter path. When the Sun's altitude is high, the angular length of the glitter path (i.e., in the direction the observer is facing) is *four times* the angle of the maximum wave slope α. This is because waves are inclined both toward and away from the observer (a difference of 2α), and reflection deviates the ray through double the angle of incidence.

This is illustrated in a different way in figure 68.1, and shows that the observer therefore sees glints extending through an angle of 4α. However, the angular width of the glitter is less than its "length" or vertical extent, being $4\alpha \sin \theta$, where θ is the sun's altitude, and so it clearly narrows as the solar

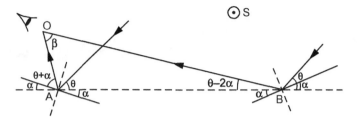

Figure 68.1. Geometry for determining the "vertical" angular extent of the glitter path

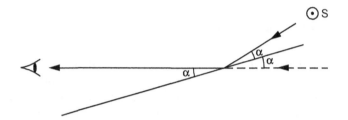

Figure 68.2. For a solar altitude of 2α, the glitter path extends all the way to the horizon.

altitude decreases. Thus, if the maximum wave angle is a shallow 5° and the Sun's altitude is 30°, the glitter path will be 20° and its width will be 10°. As can be inferred from figure 68.2, when the Sun's altitude is equal to 2α, the glitter path extends all the way to the horizon. Lower than this angle, some wave "shadowing" can occur, the glitter decreases somewhat in brightness, and the path length decreases. When the Sun's altitude is less than α, only the smaller wave slopes facing the observer are illuminated and, gradually, only the tops of the highest waves catch the sunlight and the glitter path is extinguished.

From figure 68.1, the points A and B represent respectively the nearest and farthest wavelets (or more accurately, the points of maximum slope) contributing to the glitter path. Then, just adding up the interior angles of triangle OAB we see that if β is the angle subtended at the observer's eye by the path AB,

$$\beta + (\theta - 2\alpha) + 180° - (\theta + 2\alpha) = 180°, \text{ i.e., } \beta = 4\alpha. \qquad (68.1)$$

Now we consider the *width* of the glitter path (i.e., perpendicular to AB); see figure 68.3(a–c). If the observer, at height h above the water, were looking vertically down at the glitter path, its angular width would again be determined by the maximum extent between wavelets (at P and P', say) that re-

(a)

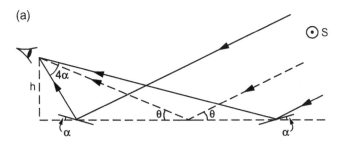

Figure 68.3 (a). The vertical extent of the glitter path

(b)

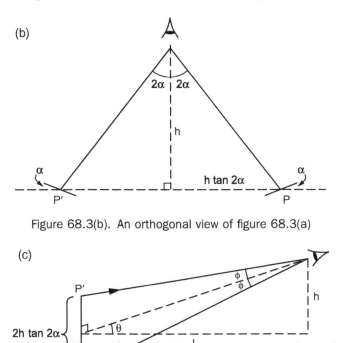

Figure 68.3(b). An orthogonal view of figure 68.3(a)

(c)

Figure 68.3(c). An oblique view of the glitter path geometry

flect sunlight into her eye. By the same arguments as above, this is again 4α. For an observer at height h but now a horizontal distance l away from the segment PP', the angular width subtended by PP' is, from figure 68.3c, 2ϕ, where

$$\tan\phi = \frac{h\tan 2\alpha}{(l^2 + h^2)^{1/2}}.$$

(68.2)

Therefore the, ratio r of angular width to angular length of the glitter path is

$$r = \frac{2\phi}{4\alpha} = \frac{\arctan\left[h(l^2 + h^2)^{-1/2}\tan 2\alpha\right]}{2\alpha} \approx \frac{h}{(l^2 + h^2)^{1/2}} \equiv \sin\theta \qquad (68.3)$$

if α is small (less than $10°$ say), as it will usually be. Equivalently then, the angular width of the glitter path is

$$2\phi = 4\alpha\sin\theta \leq 4\alpha. \qquad (68.4)$$

Consequently, if the observer is high above the water (such that $h \gg l$), then $\sin\theta \approx 1$ and the patch is close to circular. For very oblique observations, $l \gg h$, so that the path is very elongated in shape.

Q.69:   What is the path of wave intersections?

Walking by the water one morning, I noticed a single duck sitting peacefully about thirty yards from me. As it heard the sound of my approaching footsteps, it scrambled to "walk on water," flapping its wings to achieve lift as it raced out across the watery runway. Each time its webbed foot touched the water surface, of course, waves were generated. Long before it finally became airborne, a line of these waves started to interfere with each other and produce fascinating intersection patterns. Oh, how I wish I had brought my camera with me! But it did prompt the above question in my mind. If two pebbles are thrown into a pond one after the other (therefore acting as distinct "point" sources of waves), what is the locus of the point(s) of intersection of the waves? Though it's difficult to see in practice, the mathematics below shows that the path is a surprisingly well-known curve.

The points of intersection are a distance $|F'P| = r_1(t)$ away from the center of circle 1 and a distance $|FP| = r_2(t)$ away from the center of circle 2. Furthermore, if the speed of the waves is constant, then

$$\frac{dr_1}{dt} = \frac{dr_2}{dt}, \text{ i.e., } r_1 - r_2 = \text{constant.} \qquad (69.1)$$

This is just the condition that the points of intersection follow a hyperbolic path, because a hyperbola is defined as follows: given two distinct points (the foci), a hyperbola is the locus of points such that the difference between the distance to each focus is constant (figure 69.1). The resulting intersections for the two sets of waves are shown in figure 69.2.

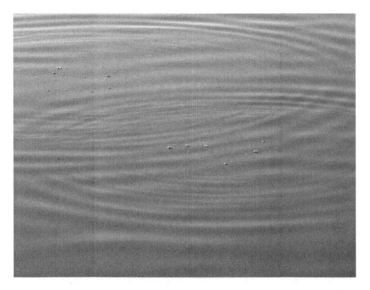

Question 69. Intersecting circular waves on the surface of a pond. The locus of the wave intersections is a hyperbola.

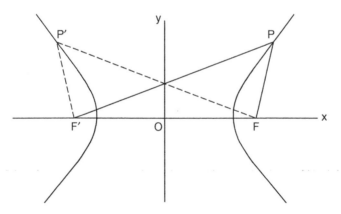

Figure 69.1. Illustration of definition (69.1)

Neat! I'm sure the duck knew that.

Q.70: How fast do waves move on the surface of water?

As they say, "it all depends . . ." I notice several types of water waves when I make my neighborhood walk in the mornings, especially if it has rained

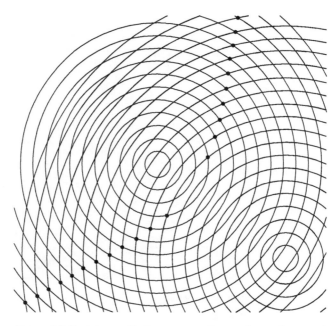

Figure 69.2. Intersecting waves on the surface of a pond

recently. A breath of wind is enough to raise ripples on the surface of puddles, and a solitary raindrop falling from an overhead branch is sufficient to set up a fascinating set of concentric circles, propagating outward smoothly from their center. Then there are longer wind-induced waves frequently visible on the surface of the inlets of the Lafayette River; rarely is it totally calm, and even then an occasional underwater dweller will break the surface to catch a fly hovering near the surface (or possibly just to attract my attention). Frequently, a committee of ducks will launch themselves into the water as I approach them, presumably unaware that I try to avoid committees at all costs. After the initial splashes have died down, the ducks produce interacting wakes as they head away from me to more suitable gathering place across the water. But I have news for them this morning: just as they heave themselves out of the water again to continue their deliberations, a dog walking a man becomes very anxious to give his two-bit(e)s worth to the preening, quacking group. He strains at the leash, and all of a sudden, shoots like a bullet into the duckfest, scattering them like, yes, duckpins, back into the water. Never have I seen a dog look so pleased with itself . . . and the committee meeting is postponed for yet another time.

Before discussing the types of wave and wake patterns behind ships and ducks and in front of stationary objects (such as rocks or protruding twigs

Question 70. Circular gravity wave patterns on a pond surface. As they move outwards, they leave an expanding region of calm water.

stuck in the streambed—see questions 71 and 72), we need to discuss some general features of waves on the surface of water, or *surface gravity waves*. These are waves propagating under the combined effects of gravity and surface tension. Frequently, one of these effects (or restoring forces) dominates the other, in which case the name is based on which force is primarily responsible—gravity or surface tension. Because *capillarity* is a phenomenon associated with surface tension (as when the adhesion forces between water and glass cause water to rise up inside a thin glass tube), the latter waves are sometimes called capillarity or capillary waves. Both types are readily observable in nature.

For the combined effects of both forces, the speed of an individual wave crest propagating in one dimension is

$$c = \lambda v = \left[\left(\frac{g\lambda}{2\pi} + \frac{2\pi\gamma}{\rho\lambda} \right) \tanh \left(\frac{2\pi h}{\lambda} \right) \right]^{1/2}, \qquad (70.1)$$

where λ = wavelength, v = wave frequency, ρ = fluid density; and h and γ are the fluid (water!) depth and coefficient of surface tension, respectively. The gravitational acceleration has magnitude g. In another formulation, the speed c of an *individual* wave of wavelength λ is given by $c = \omega/k$ and the speed c_g of a *group* of waves is $c_g = d\omega/dk$, because equation (70.1) can be expressed in terms of the *angular* frequency ω ($= 2\pi v$) and *wavenumber* k ($= 2\pi/\lambda$) as

$$\omega = ck = k\left[\left(\frac{g}{k} + \frac{\gamma k}{\rho}\right) \tanh\, kh\right]^{1/2}. \qquad (70.2)$$

It is worthwhile to consider some special cases of (70.1).

Deep Water Waves

For our purposes, "deep" here means that the wavelength is small compared with the depth of the water, i.e.,

$$\frac{h}{\lambda} \gg 1,$$

which means that

$$\tanh\frac{2\pi h}{\lambda} \approx 1.$$

Under these circumstances

$$c^2 \approx \frac{g\lambda}{2\pi} + \frac{2\pi\gamma}{\rho\lambda}, \qquad (70.3)$$

which represents the (wave speed)^2 for disturbances that "feel" the effects of gravity and surface tension, but do not "feel" the bottom of the channel, reservoir, etc. Furthermore, for "long" waves in this category, i.e.,

$$\frac{g\lambda}{2\pi} \gg \frac{2\pi\gamma}{\rho\lambda}$$

it follows that

$$c^2 \approx \frac{g\lambda}{2\pi}. \qquad (70.4)$$

Thus,

$$c \propto \sqrt{\lambda}. \qquad (70.5)$$

This is the correct relationship between velocity and wave length for ocean waves, which are completely dominated by gravity. Crudely—the longer the wavelength, the faster the wave moves.

SHALLOW WATER WAVES

At the other extreme, for "short" waves, i.e.,

$$\frac{g\lambda}{2\pi} \ll \frac{2\pi\gamma}{\rho\lambda},$$

we find that

$$c^2 \approx \frac{2\pi\gamma}{\rho\lambda}. \tag{70.6}$$

Thus,

$$c \propto \lambda^{-1/2}. \tag{70.7}$$

These waves (ripples) are completely dominated by surface tension, and the shorter they are the faster they move. Now let's go to the other extreme, and examine *shallow water waves*. This means that the depth of water is small compared with the wavelength, i.e.,

$$\frac{h}{\lambda} \ll 1.$$

Under these circumstances it may be shown from a Taylor series expansion about the origin, (i.e., a Maclaurin series) that

$$\tanh\frac{2\pi h}{\lambda} \approx \frac{2\pi h}{\lambda}.$$

These waves do "feel" the bottom. For most problems of interest in this wave context, the second term in (70.1) is negligible, so that the following result is valid for gravity waves in shallow water:

$$c^2 \approx gh. \tag{70.8}$$

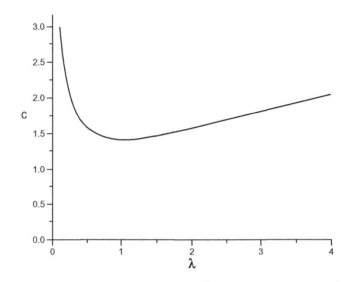

Figure 70.1. Qualitative behavior of $c(\lambda)$ based on equation (70.3)

This is an important result: it means that the wave speed is *independent of wavelength*. This implies that all the waves travel with the same speed, and any complex initial wave configuration may retain an identifiable shape for quite some time afterward. Actually, the strong inequalities ("$\ll$" and "$\gg$") we have employed to distinguish between deep water waves and their shallow water counterparts do not need to be enforced so strictly; sometimes it is sufficient to demand that $h < \lambda/2$ for shallow water waves, for example. It all depends on the context.

Let us return to equation (70.3) for deep water capillarity–gravity waves, for there is quite a bit more information we can extract. In the extreme cases given by equations (70.5) and (70.7), respectively, we have seen that the square of the speed behaves in a (i) linear and (ii) rectangular hyperbolic fashion as functions of wavelength. In the intermediate region, i.e., where the terms $g\lambda/2\pi$ and $2\pi\gamma/\rho\lambda$ are comparable, so too are both restoring forces, and the respective graphs of $c(\lambda)$ must connect. This is illustrated generically in figure 70.1 as $c \sim (\lambda + \lambda^{-1})^{1/2}$.

The speed c is clearly a minimum since $c''(\lambda) > 0$ at the critical wavelength λ_c given by the solution of

$$c'(\lambda) = 0,$$

which is

$$\lambda_c = 2\pi \sqrt{\frac{\gamma}{g\rho}}. \tag{70.9}$$

In cgs units, $\lambda_c \approx 1.72$ cm for water. For wavelengths less than or greater than this, the dominant restoring force tends to be, respectively, surface tension or gravity. If we substitute (70.9) back into (70.3) we find the corresponding *minimum* speed to be

$$c_{\min} = \sqrt[4]{\frac{4g\gamma}{\rho}} \approx 23 \ \text{cm/s} = 0.23 \ \text{m/s}. \qquad (70.10)$$

This means that any breeze or gust of wind with speed less than 0.23 m/s will not generate any propagating waves, other than a transient disturbance. Wind speeds above this minimum value will, in principle, generate two sets of waves, with wavelengths on each side of λ_c, i.e., one set with $\lambda < \lambda_c$ (capillarity waves) and one set with $\lambda > \lambda_c$ (gravity waves). Note that these results may be derived without the use of calculus: use of the arithmetic–geometric inequality gives the required result. This inequality tells us, in particular, that if $a > 0$ and $b > 0$ then

$$\frac{a+b}{2} \geq \sqrt{ab},$$

with equality occurring if and only if $a = b$. This result, which tells us that the arithmetic mean is never less than the geometric mean, is easily established by considering the inequality $(\sqrt{a} - \sqrt{b})^2 \geq 0$, and can be generalized to a set of n positive numbers, but we need only two here. Then we can recover result (70.10) by writing equation (70.3) for brevity as

$$c^2 = \alpha\lambda + \frac{\beta}{\lambda}, \quad \alpha > 0, \ \beta > 0. \qquad (70.11)$$

It follows by the above inequality that the sum of these two terms is never less than $2\sqrt{\alpha\beta} = 2\sqrt{g\gamma/\rho}$. Since the minimum of c^2 occurs when the minimum of c does, the corresponding result stated above for $c_{\min}$ is established. There is in principle no limit to the maximum speed of water waves if their wavelength is small enough. (In practice, of course, there must be.) It might be thought that a similar conclusion applies to very long waves as well, but sooner or later the waves in this limit must be considered shallow, and the maximum speed c is then just $\sqrt{gh}$ as we have seen above.

It is appropriate at this point to add a comment on wave refraction: we now have a simple model explaining why ocean waves line up parallel to the beach, even if far out to sea they are approaching it obliquely. Fix the wavelength of any particular wave you are interested in. Far out, the wave is in deep water ($\lambda \ll h_{\text{deep}}$) and so $c \propto \sqrt{\lambda}$. Nearer in, the wave is in shallow water

Question 71. Ship wave patterns seen from high above a lake in Yellowstone National Park.

$(\lambda \gg h_{\text{shallow}})$ and so $c \propto \sqrt{h_{\text{shallow}}}$, which is of course smaller than $\sqrt{\lambda}$. So that part of the wavefront nearest the beach slows down compared to that further out, and the whole thing tends to "slew" around.

Q.71: How do moving ships produce that wave pattern?

After a heavy object, such as a boulder or duck (or calculus text) is "launched" into a pond, a very complicated pattern emerges at first, one that we will not attempt to describe here. Instead, we'll focus our attention initially on the waves that emerge from the complex initial disturbance caused by the impact on the water's surface, or by a moving source such as a ship or a duck. The distinction between the speeds of individual waves and groups of them has already been made above, but as a reminder, note that if the effects of surface tension are negligible (i.e., for wavelengths longer than a few centimeters) equation (70.1) simplifies to

$$c(\lambda) = \lambda v(\lambda) = \left(\frac{\lambda}{2\pi}\right)(2\pi v) \equiv \frac{\omega}{k} = \left(\frac{g\lambda}{2\pi}\tanh\frac{2\pi h}{\lambda}\right)^{1/2}, \qquad (71.1)$$

v being the wave frequency (or reciprocal of the wave period), ω being the angular frequency of the wave and k being the wavenumber. This expression is called the *dispersion relation*, because it describes how fast waves spread out or disperse according to their wavelengths. The speed c of an individual wave of wavelength λ is now given by equation (71.1). There is an interesting relationship between c, λ, and the speed c_g of a group of waves:

$$c_g = c - \lambda \frac{dc}{d\lambda}. \tag{71.2}$$

Its derivation is left as an exercise below, and some graphical implications of this equation will also be discussed, based on figure 71.1.

This means that, given a graph of $c(\lambda)$, we can find the group speed for a particular wave speed c_0 directly from the graph. Here's how we do it: draw the tangent line to the $c(\lambda)$ graph at the point of interest, namely $(c_0(\lambda_0), \lambda_0)$ and read off the c-intercept by extending the line back to the c-axis, where, of course, $\lambda = 0$. From equation (71.2) above, this value of c is just c_g.

Exercise: Derive equation (71.2). [*Hint*: $c_g = d\omega/dk$, where $\omega = 2\pi\nu$ and $k = 2\pi/\lambda$.]

There is something else worth noting from equation (71.2), and, of course, it follows from the tangent line construction: if $dc/d\lambda > 0$, $c_g < c$, whereas if $dc/d\lambda < 0$, $c_g > c$. In the first case (such as for surface gravity waves) the speed of individual waves of a given wavelength is greater than the speed of a group of them (or, more precisely, the *envelope* of a group of them). An example of the other situation is found in capillary waves, for which the group speed exceeds the wave speed. In fact, for gravity waves, the wave speed is exactly *twice* the group speed (as noted below), while for capillary waves it is *two-thirds* of the group speed. In optics, these situations are referred to as *normal* and *anomalous* dispersion, respectively, and this is why the coordinates (c_0, λ_0) are labeled instead with subscripts n and a respectively in figure 71.1.

We have already noted two useful extremes in equations (70.8) and (70.4): for long waves $c \approx \sqrt{gh}$, i.e., the wave speed is independent of wavelength, and so there is no dispersion: waves travel with the same speed regardless of wavelength. It follows also from (71.2) that $c_g = c$. In fact, tsunamis fall into this category of shallow water waves, but here we shall focus on the other extreme, namely short waves. Then $c \approx \sqrt{g\lambda/2\pi}$, so the longer the wave, the faster it moves. This time the group speed $c_g = c/2$ as is readily shown using (71.2).

This result is extremely important in discussing the wave patterns made by ships (and ducks). To set the scene, consider a group of surface gravity waves G that after a time t has moved out a distance r from the source of the "splash," where $c_g = r/t$. Now suppose that we direct our gaze to follow a

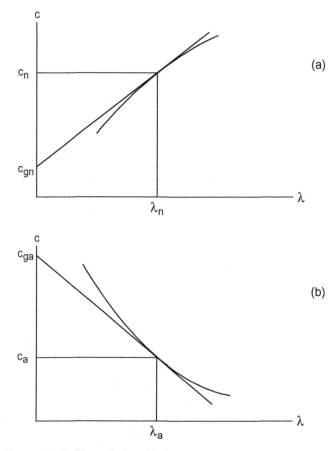

Figure 71.1. The relationship between wave and group speed

particular *wave crest* as it moves outward from the initially disturbed water. Since it travels at twice the group speed it will enter and pass though various wave groups in its journey radially outward, so when it encounters the group G, this wave crest has (approximately) the wave speed $c = 2r/t$, but from elementary calculus this is equal to the derivative dr/dt, so the governing differential equation for the position of the crest as a function of time is very simple, namely

$$\frac{dr}{dt} = 2\frac{r}{t},$$ (71.3)

the solution of which is

$$r = Ct^2$$ (71.4)

(C being a constant, or at least a quantity independent of time). (We cannot determine from this analysis whether or not it will be a different constant for different wave crests.) Equation (71.4) tells us that the radial distance covered by an emerging wave crest increases as the square of the time elapsed, and the speed of the wave is given by $r/t = Ct$, so the crest must accelerate as it moves outward; also, the area of the circular wave crest increases as the *fourth* power of time!

Our next task is to see how successive wave crests combine to form patterns if the original disturbance is now considered to be a *moving source* of waves, such as a ship or a duck. We suppose that the (very small) ship is moving at a constant speed along an imaginary x-axis from left to right, and that at this moment we set $t_0 = 0$, and place the ship at the origin of the Cartesian coordinate system, so that earlier time intervals correspond to $t_1 = -1$, $t_2 = -2$, . . . , $t_n = -n$, $n = 1$, 2, 3, . . . , etc., representing the number of time units in the past. The ship therefore travels a unit distance in a unit of our time here, so its speed is 1. When $n = 0$ (now!), the waves have had no time to spread out, so they all have radius $r = 0$. However, waves generated at all previous times, $t_n = -n$, have expanded into circles with radii proportional to n^2, as indicated by equation (71.4) above. Of course, we are simplifying things by considering discrete intervals of time, whereas the ship is producing a continuous stream of waves as it moves, but the general pattern produced will be essentially the same, and give rise to the same wave envelope.

How are we to represent the equations of these individual circles (corresponding to a particular wave crest as before)? It is not hard to see that the circle produced at $t = -n$, when the ship's bow was at the point $(-n, 0)$ has the equation

$$(x + n)^2 + y^2 = (Cn^2)^2 = An^4, \tag{71.5}$$

where A is a scaling constant that will be chosen below to elucidate the spatial development of the pattern. The overall impression is that of an arrowhead.

Looking at figure 71.2, one can just about make out the concentration of circular arcs for higher values of n, forming the back of the arrowhead (for the choice of A). For even higher values of n, the wave crests are found *ahead* of the wake formed by lower n-values, because they are, in fact, traveling faster than the ship (or supercharged duck; one such crest is indicated in the figure). Further analysis on the wake apex angle and the corresponding "side arm" waves can be found in Adam (2006).

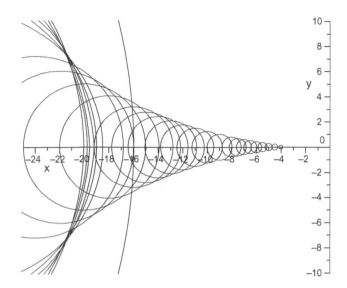

Figure 71.2. An envelope of surface gravity waves produced by a moving point source

Q.72: How do rocks in a flowing stream produce different patterns?

If we now examine very short waves or ripples such that h/λ is large, the wave speed corresponding to equation (70.6) is now

$$c = \sqrt{\frac{2\pi\gamma}{\rho\lambda}}, \tag{72.1}$$

where γ is the coefficient of surface tension and ρ is the water density. Since in this approximation c is inversely proportional to the square root of λ, shorter waves travel faster than longer waves. Using equation (71.2) it is easy to show that the group speed exceeds the wave speed, in fact, $c_g = 3c/2$, so individual ripples get overtaken by groups of this type of *capillarity* wave. We can adapt the selfsame argument used above for gravity waves to this new class of wave, driven not by gravity, but by surface tension. If an individual ripple has radius r at time t, it is part of a group whose speed is (on average) r/t, but the ripple's speed is only 2/3 of the group speed, and is represented by the derivative dr/dt, so that now

$$\frac{dr}{dt} = \frac{2r}{3t}, \tag{72.2}$$

Question 72. Capillary waves (ripples) formed upstream of a rock (stream moving from left to right).

which has solution

$$r = Bt^{2/3}, \tag{72.3}$$

B being another constant, generally a function of wavelength. For such ripples, the circle produced at $t = -n$, when the ship's bow was at the point $(-n, 0)$, has the equation

$$(x + n)^2 + y^2 = (Bt^{2/3})^2 = Dn^{4/3}. \tag{72.4}$$

Compared with duck waves, the envelope of these waves is very different; it's a smooth, rounded curve, and qualitatively describes the wave pattern *ahead* of, say, a rock or even a fishing line in a stream. The usual pattern of ship waves is often present behind these obstacles as well, being formed by the longer wavelength gravity waves described earlier. Without loss of generality, we may gain insight into this pattern by considering again our traveling point source of waves (or, if you wish, the rock moving through still water). From equation (72.2) the speed of the crest is $2r/3t$, or, using (72.3), $2B/3t^{1/3}$, a quantity that obviously decreases with time. From our discussion of surface gravity waves we note that the speed of the ship, duck, or fishing line is 1 unit distance per unit time, i.e., $v = 1$. Therefore, when $2B/3t^{1/3} \leq 1$, or $t \geq (2B/3)^3 \equiv t^*$, the ripples from succeeding disturbances reinforce just

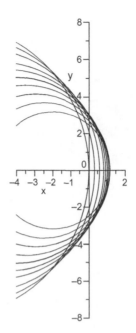

Figure 72.1. An envelope of capillary waves produced by a moving point source

ahead and behind the ship (at $t = t^*$ the ripples speed is equal to that of the ship), as indicated in figure 72.1.

Q.73:   Can waves be stopped by opposing streams?

It would certainly appear so, and the answer is yes, but let's try and make this a little more quantitative. Suppose water is flowing in the positive x-direction over an uneven streambed, and, in particular, the depth goes from deep (H_1) to shallow (h) to deep (H_2; this need not occur in a symmetrical manner). If the extremes are great, then from the conservation of linear fluid momentum we can say that the stream speed will be much less in the deep regions than in the shallow, so much so for really deep regions that we can effectively consider it to be zero for the present purposes. In the shallow region, the stream speed is $V > 0$. Suppose also that waves with speed $c_0 > 0$ in the (still) deep water move into the shallow (flowing) region; how will the speed of the waves change as this happens? If the wavelength in deep water is λ_0, the frequency of the waves is $\nu_0 = c_0/\lambda_0$, and this will also be their frequency in the deep water on the far side of the shallows. If the speed of the waves in the shallows relative to the stream is c, then relative to the still water (or a stationary observer) it is $c + V$; if their wavelength is now λ then their frequency is now $(c + V)/\lambda$. If

Question 73. Waves formed on the surface of a stream moving from right to left.

waves are neither created nor destroyed between these two places, then the same number of waves must pass each point, which means that the frequencies are identical. Thus,

$$\frac{c_0}{\lambda_0} = \frac{c+V}{\lambda}. \tag{73.1}$$

At this point, we can consider two options: the first is the case for which all of the depths H_1, H_2, and h are large compared with both λ_0 and λ; then we have from equation (70.4) that

$$\lambda_0 = \frac{2\pi}{g} c_0^2 \text{ and } \lambda = \frac{2\pi}{g} c^2. \tag{73.2}$$

Upon substituting these expressions into equation (73.1) we find, after a little rearrangement, the following quadratic equation for the desired speed c:

$$c^2 - cc_0 - Vc_0 = 0. \tag{73.3}$$

This has roots

$$c = \frac{1}{2} c_0 \left(1 \pm \sqrt{1 + \frac{4V}{c_0}}\right), \tag{73.4a}$$

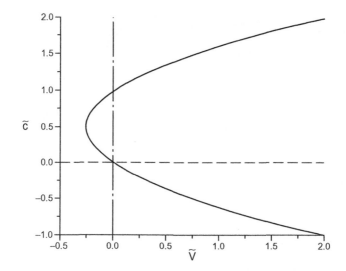

Figure 73.1. The wave speed $\tilde{c}$ as a function of the stream speed $\tilde{V}$

or in terms of the dimensionless quantities $\tilde{c} = c/c_0$ and $\tilde{V} = V/c_0$,

$$\tilde{c} = \frac{1}{2}\left(1 \pm \sqrt{1 + 4\tilde{V}}\right). \tag{73.4b}$$

Does this allow for the possibility of waves moving in both directions, one for each root? Figure 73.1 shows both branches of $\tilde{c}$.

Note that the c-intercepts are just 0 and 1, and the vertex of the parabola is located at $(-1/4, 1/2)$. The portion of the graph (including the vertex) for which $\tilde{V} < 0$ and $0 < \tilde{c} < 1$ corresponds to waves moving *against* the stream; clearly, here in dimensional terms $c < c_0$; the wave speed relative to the stream decreases. Note also that there is *no solution* c if $\tilde{V} < -1/4$ (or $V < -c_0/4$). This can be interpreted to mean that surface gravity waves will not be able to traverse an opposing stream if the speed of the stream exceeds $c_0/4$. At the vertex, $\tilde{c} = 1/2$, so $c = c_0/2$, and individual waves are still able to move upstream, but the group speed is half this, and wave groups at this speed are unable to advance. Individual waves move upstream through this limiting group and like old soldiers, just fade away, for there are no more groups ahead to accommodate them. But the energy associated with these wavelets surely cannot just disappear, can it? As wave groups move upstream, they cannot get past this limit point in the (V, c) parameter space, so groups moving faster than $c_0/4$ must eventually slow down and pile up there, concentrating the energy in such a way that the waves must increase in height (but notice now we are on shaky ground, because the formula we have used for wave speed is for very small amplitude waves). If this is true, then eventually they will break, and their energy will be dissipated rapidly by frictional processes not considered here. And, lo, the waves are no longer present beyond this point. (For $\tilde{V} > 0$, $c > c_0$, as indicated from the upper branch [the lower branch corresponds to negative values of $\tilde{c}$] and hence $\tilde{c}$, and by hypothesis, $c > 0$ here.)

Q.74:   How far away is the storm?

Perhaps you are spending a few days at the beach, and it's getting pretty windy out there. Suppose that surface waves are generated by a mid-Atlantic storm, and arrive at the British coastline with a period of 15 seconds. One day later, being *very* observant, you notice that the period of the arriving waves has dropped to 12.5 seconds (the wavelength has decreased, and so has the wave speed of the waves). How far away, approximately, did the storm occur (to the nearest 100 miles)?

It is helpful to think of waves of a certain period $T = 2\pi/\omega$ moving with the group speed c_g, so that waves of period T arrive at the coast from a distance d in a time $t = ?$ For ocean waves,

$$\omega = \sqrt{gk} = \sqrt{2\pi g/\lambda} \Longrightarrow T = \sqrt{2\pi\lambda/g}. \tag{74.1}$$

The energy of the storm travels at the speed

$$c_g = c/2 = \omega/2k = (\sqrt{g/k})/2 = (\sqrt{g\lambda/2\pi})/2 = [(\sqrt{g/2\pi})/2]$$
$$\times (\sqrt{g/2\pi})T = gT/4\pi. \tag{74.2}$$

Traveling at the group speed from a distance away of d miles, it takes time $t = d/c_g$ for wave groups to reach the coastline, so (74.2) implies that

$$t = 4\pi d/gT,$$

where $T = 15$ and 12.5 s, respectively. Next, equating the distance traveled for both groups, we have that

$$d = c_g t = gTt/4\pi = (32 \times 15) \times t/4\pi$$
$$\equiv (32 \times 12.5) \times (t + 24 \times 3600)/4\pi, \tag{74.3}$$

from which $t = 4.32 \times 10^5$ s. On substituting this value back into *either* expression for d we find that $d \approx 3100$ miles (or approximately 5000 km). Since

Question 75. Circular patterns formed by capillary waves moving out radially on the surface of a pond. The source of the waves is a struggling moth. Courtesy Heather Renyck.

we now know the storm hit 3100 miles from us, all our hard work in taking precautions may not have been necessary after all!

Q.75:   How fast is the calm region of that "puddle wave" expanding?

What on earth am I talking about here? Well, please consult figure 75.1. It is representative of the pattern formed when a small raindrop falls on a puddle, and the waves expand rapidly, with an expanding region of calm behind them. These waves generally fall into the capillary category because surface tension is the dominant restoring force, so we'll use those equations to discuss the question.

To simplify things, we'll assume that the puddle is deep in the sense that its depth exceeds half that of the longest wavelength that I observe on its surface. Then, including the effects of surface tension, we can use (70.2) to write

$$\omega^2 = \left(gk + \frac{\gamma}{\rho} k^3 \right) = gk + bk^3, \qquad (75.1)$$

where $b = \gamma/\rho$. Differentiating this expression implicitly, we find that the group speed of deep water gravity–capillarity waves is

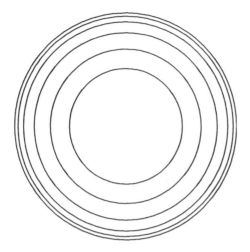

Figure 75.1. The wave pattern formed by a raindrop falling on a puddle

$$c_g(k) = \frac{d\omega}{dk} = \frac{g + 3bk^2}{2(gk + bk^3)^{1/2}}. \tag{75.2}$$

If this group speed has a minimum value for some wavenumber k, then the last wave group to move out from the center of the disturbance will travel with this speed, and since the energy also travels with the group speed, the region

$$r < (c_g)_{\min}t$$

is calm. A further differentiation shows that

$$\frac{dc_g}{dk} = 0 \text{ when } k = \left(\left[\frac{2}{\sqrt{3}} - 1\right]\frac{g\rho}{\gamma}\right)^{1/2} = k^*. \tag{75.3}$$

and, furthermore, it is readily shown by the first derivative test that this critical number defines a *minimum* of c_g. This means that there must be a region a calm water behind the wave groups, and it expands with speed given by that of the slowest wave group, i.e.,

$$(c_g)_{\min} = c_g(k^*) = \frac{\sqrt{3} - 1}{\left(\frac{2}{\sqrt{3}}\right)^{1/2}\left(\frac{2}{\sqrt{3}} - 1\right)^{1/4}}\left(\frac{g\gamma}{\rho}\right)^{1/4} \approx 1.09\left(\frac{g\gamma}{\rho}\right)^{1/4}. \tag{75.4}$$

Now it's time to put in the numbers. We use $g = 9.81 \text{ m/s}^2$, $\gamma = 0.074$ N/m, $\rho = 10^3 \text{ kg/m}^3$ to find that $(c_g)_{\min} \approx 18 \text{ cm/s}$ (or, mixing our units, about 0.4 mph). The wavelength corresponding to this minimum speed is just

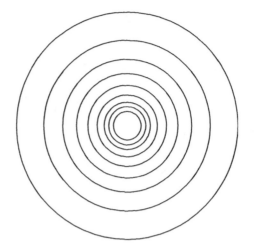

Figure 75.2. The wave pattern formed by a boulder falling on a pond

$\lambda^* = 2\pi/k^* \approx 4.4$ cm. This means that for the deep water approximation to be valid, $h \gtrsim 2$ cm, which may well have been the case (I did not measure it at the time). Even if this is not the case, and the puddle is not "deep," a similar type of result holds true.

As a final comment, figure 75.2 is the kind of pattern ultimately produced when a larger object, such as a boulder, is thrown into a pond. After the initial disturbance to the surface has settled into circular waveforms, the waves are longer than in the puddle, and are dominated by gravity, so the longer waves move out faster. This is, of course, the reverse of the behavior of capillary waves, for which the shorter waves move faster. In either case, the energy delivered to the water surface is propagated outward by the waves.

Q.76: How much energy do ocean waves have?

Speaking of wave energy, we can perform some calculations on this very topic. How much energy do they have? We'll focus first on the deep water linear surface gravity waves to simplify the mathematics, and then insert some numerical values for the waves. There is a quick, dirty, and valuable physical way to obtain this result, and also a longer but more mathematically satisfying procedure; it would be nice to adopt a bipartisan approach and do both, but in the interests of space we will retain the more physical approach here.

Imagine a thin column of water in the wave at height z above the mean level (were there no wave), with width δx and length y parallel to the crest lines (out of the page as you read this—Don't get wet!). It has gained potential energy by an amount equal to mass×gravity×height, or $\rho zy\delta x \times g \times z$; to represent the whole crest it must be integrated over the whole length $\lambda/2$. Thus, if the wave profile is sinusoidal (as linear theory demands), then $z = A \sin kx$, A being the wave amplitude, which we'll take as real for now, and the potential energy in one wavelength is

$$\int_0^{\lambda/2} \rho gyA^2 \sin^2 kx\, dx = \frac{1}{2}\rho gyA^2 \int_0^{\lambda/2} (1 - \cos 2kx)\, dx = \frac{1}{4}\rho gyA^2 \lambda. \quad (76.1)$$

But wait a moment—what about the rest of the wavelength? Haven't we forgotten the trough, and since that is *below* the mean surface doesn't the total potential energy cancel out to zero? That's a question worth thinking about . . . Since in linear theory the troughs and crests are symmetrical, we can think of each wavelength as having been formed by scooping out the water from the flat surface (making a trough) and piling it up to make a crest, so it is not necessary to do any more calculations for the potential energy of the wave. This isn't quite what happens, of course, but I did say this was quick and dirty . . . So now all that remains is for the kinetic energy of the wave to be calculated. That's easy; just double the above result! But wait another moment—what is going on now? Nothing more than a little mathematical sleight of hand, perfectly acceptable, but it needs a little more clarification. It is well known that two equal linear wavetrains are equivalent to a standing wave, because, for example,

$$A \cos(kx - \omega t) + A \cos(kx + \omega t) = 2A \cos kx \cos \omega t, \quad (76.2)$$

so that at each location x the surface rises and falls in time. Less familiar is the fact that two suitably chosen standing waves are equivalent to a propagating (or progressive) wave. This follows, for example with two waves of equal amplitude one-fourth of a wavelength out of phase and one-fourth of a period out of phase, because

$$A \cos kx \cos \omega t + A \sin kx \sin \omega t = A \cos(kx - \omega t). \quad (76.3)$$

When one of these is at maximum amplitude, the other is momentarily "flat," with zero net potential energy, and all its energy in the form of particle motion. But one-fourth of a period later, these profiles are interchanged, so there must be "equipartition" between the averaged potential and kinetic energies. Hence, the total energy is $\frac{1}{2}\rho gyA^2\lambda$, or per unit surface area (dividing by $y\lambda$) it is

$$E = \frac{1}{2}\rho gA^2, \quad (76.4)$$

which is independent of the wavelength or frequency of the wave; and nowhere has the depth entered in, other than through the assumption of "deep" water. This result is standard in one sense for all waves: energy is proportional to the square of the amplitude. Continuing the calculation, for an amplitude A of 1 meter, and density of water of 1000 kg m^{-3}, the energy per square meter for such a wave is

$$E = \tfrac{1}{2} \times 10^3 \times 9.8 \times 1 = 4900 \text{ joules m}^{-2},$$

or about a tenth of the energy of a steel breaking ball when it smashes into a building. This energy is delivered at the group speed, which is half the wave speed, i.e., $c/2$. For a wavelength of, say, 100 m, the wave period T is 8 s and the group speed is about 6 m/s, so the energy per meter length of wave in the y-direction is about 30 kW (kilowatts) per meter. If we hazard a guess of about 2000 km for the coastline of Britain, this corresponds to an energy of about 6×10^4 MW (megawatts), just for this rather low frequency wave (which, of course, could not be expected to be the same all around the island), so the actual energy flow is probably much higher than this.

For an interesting energy comparison, consider the power output of an average human adult. Typically, we consume about 2500 food calories (Calories) per day, and 1 Calorie $= 4 \times 10^3$ J; since there are about 10^5 seconds in a day, our power output P is

$$P \approx 2.5 \times 10^3 \text{ Cal/day} \times \left(\frac{4 \times 10^3 \text{J}}{1 \text{ Cal}} \right) \times \left(\frac{1 \text{ day}}{10^5 \text{s}} \right) = \frac{10^7 \text{J}}{10^5 \text{s}} = 100 \text{J/s} = 100 \text{W},$$

i.e., that of a 100-W light bulb! Therefore, the above estimate of wave energy per meter, 3×10^4 W, is the equivalent of the power output of 300 people crowded together in an auditorium. Cool . . . or, rather, warm . . .

Q.77:   Does a wave raise the average depth of the water?

No, but it does raise the center of mass of the water. Let's examine these statements and their implications in more detail. Figure 77.1 shows the surface profile for a simple (and ideal) wave. Let's first examine the effect of a single wavelength on the average depth of the previously calm water. Well, by the conservation of mass, there'll be no change, will there? Correct. But what's the point of taking a calculus course if we don't use it everywhere we can? Okay,

Figure 77.1. Part of a surface undulation caused by a wave

I'm convinced. We'll take $y=0$ as the undisturbed water surface, and super-impose on it the wave "segment" of amplitude A and wavelength λ defined by

$$y(x) = A \sin \frac{2\pi x}{\lambda}. \tag{77.1}$$

We also take the depth of the undisturbed water to be H, specifically, the bottom boundary is defined by $y = -H$. The average value of the depth with the surface undulation over a wavelength is defined by

$$\bar{H} = \frac{1}{\lambda} \int_0^\lambda (H + y(x)) dx = \frac{1}{\lambda} \int_0^\lambda \left(H + A \sin \frac{2\pi x}{\lambda} \right) dx = H. \tag{77.2}$$

No surprises there. But if we break the interval of integration into two equal halves, then at least we see by how much the average depth deviates from H for the crest and trough. Thus, on $[0, \lambda/2]$,

$$\bar{H} = \frac{2}{\lambda} \int_0^{\lambda/2} \left(H + A \sin \frac{2\pi x}{\lambda} \right) dx = H + \frac{2A}{\pi}, \tag{77.3a}$$

and similarly, on $[\lambda/2, \lambda]$

$$\bar{H} = \frac{2}{\lambda} \int_{\lambda/2}^\lambda \left(H + A \sin \frac{2\pi x}{\lambda} \right) dx = H - \frac{2A}{\pi}. \tag{77.3b}$$

Therefore, the crest increases the average water depth by $2A/\lambda \approx 0.64A$ and the trough decreases it by the same amount. However, because of the amount of water *added* to the previously flat surface that under the crest is more than that *left* under the trough, so the height of the center of mass of the body of water *does* change. To justify the first statement, we can just examine the graph of the sine function (77.1) above . . .

The height of center of mass $\bar{y}$ of the water above the bottom is over one wavelength,

$$\tilde{H} = \frac{\frac{1}{2} \int_0^\lambda (H + y(x))^2 dx}{\int_0^\lambda (H + y(x)) dx} = \frac{\frac{1}{2} \int_0^\lambda (H + A \sin \frac{2\pi x}{\lambda})^2 dx}{\int_0^\lambda (H + A \sin \frac{2\pi x}{\lambda}) dx} = H + \frac{A^2}{2H} > H, \tag{77.4}$$

after a little manipulation (if you know your trigonometric identities!)

Question 78. A ship wake used to determine the approximate radius of the Earth. Courtesy David Lynch.

Exercise: Establish the result (77.4).

Q.78:   How can ship wakes prove the Earth is "round"?

Based on a photograph of the turbulent wake of a ship on which he was travelling in the Indian Ocean, a friend of mine (Dave Lynch) was able to deduce that the Earth is round and to estimate its radius (though I think that he probably knew it was round before doing his calculations). How did he accomplish this? Cleverly, yet simply is one answer, and the details of the way he did it are set out here (along with the photograph he took).

If the Earth were flat and the horizon infinitely far away, then the parallel edges of a long, straight wake would appear to converge to a point on the horizon (neglecting any atmospheric diminution of the distant path), an obvious effect of the observer's perspective. Of course, this would not be the case for a flat but finite Earth (e.g., a circular disk), this possibility will not be considered here; after all, Dave's ship didn't fall off the edge, but then again, maybe it didn't travel far enough . . .

However, as the photograph indicates, the width of the wake is not zero at the observable horizon, indicating a finite distance to the horizon, because of the surface of the earth continually "falling away" below the horizontal tangent plane of the observer. If this same effect were to be observed at $90°$

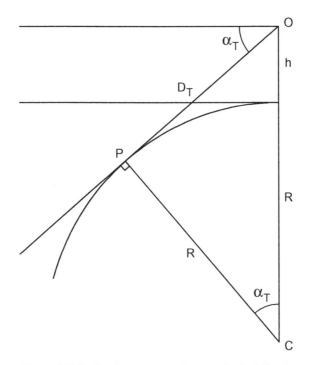

Figure 78.1. Basic geometry for a spherical Earth

to the original direction, this would indicate that we abide on a spheroidal planet rather than a cylindrical one! From here on out, we base our calculations making the assumption of a *spherical Earth*.

Suppose that the observer (Dave) is at a height h above the surface of the Earth (in this case, the ocean), or, more precisely, his eyes are. The length $D_T = OP$ is the distance to the horizon as seen by the observer at O (see figure 78.1) and α_T is known by navigators as the angle of dip. It is, as its name implies, the angle below the horizontal made by the line from the observer to the horizon.

If R is the radius of the Earth, then application of Pythagoras' theorem yields the relationship

$$D_T = (2Rh + h^2)^{1/2}. \tag{78.1}$$

(We have encountered this result before; do you recall where?) From the geometry of the figure, another useful result is

$$\sin \alpha_T = \frac{D_T}{R + h}. \tag{78.2}$$

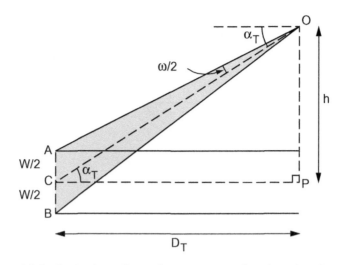

Figure 78.2. Basic three-dimensional geometry for observing the wake

Figure 78.2 depicts the three-dimensional geometry for an observer directly above the center line PC of the wake of width W. The angle subtended at the observer's eyes by the width of the wake at the horizon is ω.

From these figures it can be shown that

$$\omega = 2 \arctan \left[\frac{1}{2} \frac{W}{(D_T^2 + h^2)^{1/2}} \right]. \tag{78.3}$$

We will put some numbers into some of these expressions in due course, but two reasonable approximations render this last result much simpler, in practice. Since ω is a very tiny angle here and the ratio h/D_T is also tiny, equation (78.3) above reduces to the approximation used by Dave, namely,

$$\omega \approx \frac{W}{D_T}. \tag{78.4}$$

What is really needed now is to express what we want, R, in terms of what can be measured from the photograph, i.e., ω, W and h. From equations (78.1) and (78.4) we find that

$$R = \frac{W^2 - \omega^2 h^2}{2h\omega^2}. \tag{78.5}$$

Since mathematicians like to generalize, here is another exercise for the reader: without using the above approximations, show that

$$R = \frac{W^2}{8h} \cot^2 \frac{\omega}{2} - h. \tag{78.6}$$

How does the estimated radius of the Earth, R, vary (in particular) with W and h? Clearly, whether we use equation (78.5) or (78.6) it must be the case that $(\partial R / \partial W) > 0$, i.e., an overestimate of the wake width increases the estimate of R. Again, whether one uses the approximation (78.5) or the exact result (78.6), it is the case that $(\partial R / \partial h) < 0$, so that an overestimate of the observer's altitude h leads to an *underestimate* of R.

Now it's time to put in some numbers, courtesy of Dave (details can be found in the *Applied Optics* article cited in the references). He uses $W = 30$ m, $h = 32$ m, and $\omega \approx 4.6$ minutes of arc (recall that there are sixty minutes of arc in one degree). There is no need to use anything but equation (78.5), from which we find $R \approx 7850$ km, which compares fairly well with the actual value of 6380 km, given the difficulty of making accurate measurements from a photograph.

In the forest

Q.79:   How high can trees grow?

The mechanical or elastic properties of trees determine whether or not a tree trunk will buckle under its own weight, and whether or not a branch will bend or break under a load. Thus, stiffness and strength are properties determined by the elasticity of the structure. Consider some facts concerning trees: (i) In relation to its density, wood is stiffer and stronger, both in bending and twisting, than concrete, cast iron, aluminum alloy, or steel; (ii) trees are frugal in their use of resources for growth; (iii) they "use" the "principle of minimum weight"; this self-explanatory principle means that because every material used to stiffen or support a structure also adds to the total load the structure must bear, adding height (i.e., as in growth) must involve a trade-off between the strength of a structure and its weight (as noted in chapter 3, this is why King Kong and giant ants will not take over the world, for they cannot exist). This also explains, in part, at least, why the Earth's tallest, most massive, and longest living organisms are trees. Thus, the largest living animal, the blue whale, rarely exceeds 110 feet in length and 180 tons in weight, whereas giant sequoias (*Sequoiadendron giganteum!*) have grown to heights of over 310 feet (94 meters) with girths of as much as 100 feet (30 meters). California's General Sherman sequoia is over 270 feet high and estimated to weigh at least 2500 tons. The very tallest trees are the coast redwoods (*Sequoia sempervirens*), though their girth is considerably smaller than that of sequoias. Some of the oldest trees are more than 2000 years old, the oldest bristlecone pine (*Pinus longaeva*) being perhaps more than 4700 years of age! If left undisturbed, trees will stand for many decades after their death.

What follows then, is a discussion based on purely mechanical ideas; it will be limited, therefore, by ignoring one (if not *the*) major aspect of tree growth, namely that trees are limited by the height to which nutrient-rich water can be drawn by the tree (via transpiration). However, we shall content ourselves

Question 79. Pine trees in an Alberta wood.

here with an equally interesting problem in mechanical engineering: *the buckling of a uniform column* (our tree) *under its own weight.* By means of mechanical arguments (summarized at the end of this question), and also from a study of the relevant differential equations, it can be established that the maximum height H_{max} of a uniform cylinder radius R_{max} is proportional to $R_{\mathrm{max}}^{2/3}$, or, as perhaps is more readily appreciated, the radius R of our "tree" is proportional to $H_{\mathrm{max}}^{3/2}$. (Both approaches are discussed in Adam (2006).) Thus, as H increases over time (up to H_{max}), R *increases at a greater rate*; if the height doubles, the radius increases by a factor of $2^{3/2} \approx 2.8$. It can be argued, of course, that a tree is not merely (or even) a vertical cylinder, because of the complex branch structure, the leaf canopy, and the fact that some trees taper toward the top. But this simple model is surprisingly useful for any species of tree, given knowledge of the constant of proportionality k in the equation for H_{max}:

$$H_{max} = k\,(R_{max})^{2/3}. \tag{79.1}$$

The constant k depends on the mechanical properties of each particular tree (such as density, and the linear elasticity, or stress-to-strain ratio); in the short table below H_{max} has been calculated for ash, black oak, and redwood trees:

Tree (and estimate of typical R)	H_{max} (m)	H_{max} (ft)
Ash ($R_{max} = 9$ in $= 0.23$ m)	60.8	200
Black oak ($R_{max} = 18$ in $= 0.46$ m)	90.2	296
Redwood ($R_{max} = 1$ m)	164.0	538

Clearly, according to this model, trees are built with a large safety factor! The tallest known redwood tree, Hyperion, a California coastal redwood, is 379 ft high. The corresponding records for some other trees are Douglas fir (302 ft), giant sequoia (General Grant, 310 ft), ponderosa pine (223 ft), cedar (219 ft), Sitka spruce (216 ft), and beech (161 ft). Typical values for the black oak and white ash are 50–80 ft (diameter 1–$2\frac{1}{2}$ ft) and 80 ft (diameter about 2 ft), respectively. The redwood is typically about 350 feet high.

This upper limit H_{max} has been derived purely on the basis of mechanical considerations. As noted at the beginning of this section, a related, but different constraint, totally ignored here, is that of the ability of the tree to move water through the trunk to the branches and leaves. This is done by transpiration, and clearly there is a limit to how high a tree can be for this mechanism to work efficiently. This will provide an upper limit *less* than H_{max} (I suspect), and hence a more realistic one. Engineers like to build in safety factors as well! Furthermore, it is interesting that on the basis of a purely engineering type of analysis we have been able to deduce the basic observations concerning the relationship of tree radius to tree height. A further point of interest is the suggestion that this model may also be appropriate in animal models, since most body segments are roughly cylindrical and may be built to withstand buckling.

DIMENSIONAL ARGUMENTS AND MECHANICAL "DERIVATION" OF (79.1)

If the ratio of the height to radius of a self-supporting vertical column is too large, it will become unstable to lateral deflections, as seems eminently rea-

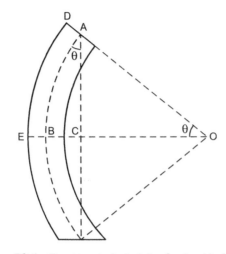

Figure 79.1. Exaggerated sketch of a buckled beam

sonable to anyone who has experimented with flexible rods. The reason is that if there is a "bend" in the previously vertical beam, then there is a torque (or moment tending to cause rotation) due to the weight of the column above the point of maximum sideways deflection. In equilibrium, this is balanced by the respective compressive and tensile stresses on either side of the column. When this elastic torque (τ_e) is not sufficient to balance the so-called gravitational torque (τ_g), the column buckles. Thus, for stability

$$\tau_e > \tau_g. \tag{79.2}$$

We wish to translate this inequality into one explicitly relating the height H of the vertical column to its diameter $D = 2R$. In what follows, note that the expression $a \sim b$ is an approximating phrase meaning "a behaves like b." From figure 79.1 it follows that if W is the weight of the column, $\tau_g \sim Wa$ (a being the total lateral deflection and ignoring numerical constants) and the stresses σ give rise to forces $\sim \sigma D^2$ and moments $\tau_e \sim \sigma D^3$. From *Hooke's law* (oversimplification: stress is proportional to strain) it follows that the maximum stress

$$\sigma_{\max} \sim E\left(\frac{\Delta H_{\max}}{H}\right), \tag{79.3}$$

where E is Young's modulus and $\Delta H_{\max}$ is the maximum amount that the sides can be stretched and compressed, respectively.

Thus, the inequality (79.2) becomes

$$Wa < E\left(\frac{\Delta H_{\max}}{H}\right)D^3. \tag{79.4}$$

From the geometry of figure 79.1, if r is the (large) radius of the approximately circular arc of the neutral line AB, then

$$H = 2\theta r \text{ and } H + \Delta H = 2\theta \left(r + \frac{D}{2} \right),$$

so that

$$\frac{\Delta H}{H} = \frac{\theta D}{H} \sim \left(\frac{aD}{H^2} \right) \tag{79.5}$$

since $\sin \theta \sim \theta$ and hence $\theta \sim 2a/H$. From equations (79.4) and (79.5), therefore,

$$Wa < E \left(\frac{aD}{H^2} \right) D^3,$$

or

$$H < \left(\frac{EHD^4}{W} \right)^{1/3} = \left(\frac{E}{\rho} \right)^{1/3} D^{2/3} = H_{\text{max}}, \tag{79.6}$$

where the fact that $W \sim HD^2 \rho$ has been used, ρ being the density of the column. Remember that this is essentially a dimensional argument, but it is valid to within multiplicative constants of order unity. Identifying $k = (4E/\rho)^{1/3}$ establishes the upper bound for H_{max} in equation (79.1).

Finally, a simpler argument can be based on the reasonable assumption that the pressure remains constant as the tree grows. This assumption is more likely to be valid for *bone* growth, in fact, but it is included here for completeness. Pressure is defined as force per unit area, and for a vertical column, the force acting on any horizontal cross section is the weight of the column above it. Weight is proportional to volume, which is in turn proportional to (size)^3, so the pressure is proportional to the quantity H^3/R^2. If the pressure is to remain constant as the tree (or bone) grows, then if this assumption is valid, the height of the tree (at any stage of its growth, in fact) is proportional to the two-thirds power of its radius.

Q.80:   **How much shade does a layer of leaves provide for the layer below?**

This is an important question because, ideally, parts of all leaves should be in sunlight for the maximum benefits of photosynthesis to be acquired. Al-

Question 80. Leaf shadows. Notice how effective each tree is at blocking out sunlight.

though leaves are generally not circular in shape, that is not a significant issue, because we are concerned here with the largest circular area of the leaf that completely blocks off light. In practical terms, then, we are dealing with an *effective* diameter, i.e., the diameter of the largest circle that can be fully inscribed within the perimeter of the leaf. Thus, a thin, elongated pine leaf will generally have a smaller effective diameter than a round or elliptical leaf of equal area. Also, leaves are not entirely opaque, but typically they absorb about 95% of the light incident upon them, so leaves act as light filters. Let us take circular leaves of effective radius r and "scatter" them randomly in n layers with an average density (i.e., number) of ρ leaves per unit area of ground. This means that this density ρ is uniformly distributed vertically among n layers, and, furthermore, we assume for simplicity that the leaves are spaced within each layer so that none overlap. This is not completely justifiable, of course, but it does provide us with an upper bound on the area intercepted by the totality of leaves.

Now we define the projection of leaves as the proportion of ground covered by the leaves' shadows when the leaves are illuminated by sunlight on a cloudless day; it is thus the proportion of the light flux that they intercept. The average projection of each of the n layers is therefore $\rho\pi r^2/n$. If the layers are far enough apart (in terms of leaf diameter, i.e., in the sense described above), then a proportion $1 - \rho\pi r^2/n$ of the incident light gets through the first layer, and of that, the same proportion $1 - \rho\pi r^2/n$ is transmitted, and so on down to the nth layer. The proportion P_n of incident light penetrating all n layers (multilayers) is thus

$$P_n = \left(1 - \frac{\rho \pi r^2}{n}\right)^n. \tag{80.1}$$

At this point there are two limiting cases or extremes of interest. The first is the case of the *monolayer* for which $n = 1$ (this is not uncommon, e.g., some types of palm tree); the second is the limiting case of $n \to \infty$, for which, of course,

$$P_\infty = \lim_{n \to \infty} P_n = \lim_{n \to \infty} \left(1 - \frac{\rho \pi r^2}{n}\right)^n = e^{-\rho \pi r^2}. \tag{80.2}$$

Thus, in assuming that the leaves are distributed evenly among the n layers (and infinity is a whole *lot* of leaves, to be sure) and that the layers are far enough apart to be independent of each other, we have arrived at a well-known result for the exponential decay of the incident light as it passes through these leafy filters. We shall see this type of result again when we examine the opaqueness of a wood in question 82. Note that for any finite number of leaves, there is upper bound to n; one cannot have fewer than one leaf per layer! The implications of the limiting form (80.2) are not surprising: the larger the density of leaves per unit area (ρ), and also the larger the average effective radius r of the leaves, the more shade there will be on the ground (or at any given level). A more subtle implication is that for two trees with the same average leaf area, but with very differently shaped leaves, the tree with the most elongated leaves (and hence smaller value of r) will allow most light to reach the ground.

Q.81: What is the "murmur of the forest"?

When wind blows across a stretched wire (i.e., perpendicular to it), notes or humming sounds may be produced. These sounds are called *aeolian* tones; an aeolian harp is a simple musical instrument that emits sounds when wind "plays" it, as a result of *vortex shedding*—a repeated pattern of vortices alternately separating from each side of an obstacle. It is not stretching the imagination, therefore, to realize that pine needles, leafless twigs, and small branches can all, by the same principle and under appropriate conditions, produce aeolian tones . . . so trees have voices, characteristic of their species. Thus, the relatively large twigs and branches of the oak tree may be expected to produce many low tones compared with innumerable fine needles of the pine tree, which will produce a range of higher-pitched notes. How do such notes blend together? How are the loudness and pitch resulting from the symphony of combined notes related to the individual properties of each "musician"?

We can make a simple excursion toward answering these questions. Consider the result of adding two sinusoidal wave trains of equal amplitude a, angular frequencies ω_m, ω_n, and phases δ_m, δ_n respectively. Then

$$a[\cos(\omega_m t - \delta_m) + \cos(\omega_n t - \delta_n)] = 2a\cos\tfrac{1}{2}[(\omega_m + \omega_n)t - \delta_m - \delta_n]$$
$$\times \cos\tfrac{1}{2}[(\omega_m - \omega_n)t - \delta_m + \delta_n].$$
(81.1)

As is well known, if $|\omega_m - \omega_n|$ is small enough, then this describes the phenomenon of "beats," where an oscillation with frequency and phase exactly equal to the means for the two original oscillations (the first cosine on the right) is modulated by a slowly varying "carrier wave" (the second cosine on the right). In general, however, we can see that the higher frequency part (if $\omega_m > 0$ and $\omega_n > 0$) has frequency depending on the sum of ω_m and ω_n, whereas the lower frequency part depends on the difference. For an example, if two sound waves with frequencies of 256 and 254 Hz are played simultaneously, a beat frequency of 2 Hz will be heard (the human ear is capable of detecting beats with frequencies of 7 Hz and below). Generally, the combination "note" will have a pitch intermediate between its components. Many such sounds would be expected to merge into a "quasinote" whose pitch is the approximate average of those of the individual components. Thus, as Humphreys points out in *Physics of the Air*, it may be concluded that "the whisper of a tree . . . has substantially the same pitch as that of its individual twigs or needles; just as the hum of a swarm of bees is pitched to that of the average bee." We now examine, in an approximate manner, the *intensity* of the blended notes.

The basic idea is to consider n individual sounds of unit amplitude and the same pitch or frequency but arbitrary phase (at least for the moment). This is believed to be a reasonable approximation to the aeolian blend of a tree. If all the notes had the same phase at a given instant, the combined amplitude would be equal to n, and the combined intensity would be n^2. If instead exactly half of the notes had a given phase (if n is even) and the other half had exactly the opposite phase (i.e., 180° difference) then the combined intensity would be zero. What we do now for simplicity is to consider that the n sounds have phases that are either "+" or the exact opposite ("−"), with equal probability. This is an example of a binomial distribution (X, say), with probability of occurrence $1/2$. By confining the phases to these options only we can gain insight into the more realistic case by means of some simple mathematics. If all n sounds (or oscillations) have the same phase (say $+$) they will reinforce one another, and the resulting intensity will be n^2. If however, one of the oscillations has the $(-)$ phase, then it will cancel out one of the remaining $n-1$ $(+)$ oscillations, leaving $n-2$. Continuing in this way, we see that for *two* oscillations of $(-)$ phase, two $(+)$ sounds will be canceled, leaving $n-4$ of them, and so on. Statistically, then, the average (or *expected*) value E of the intensity, I_{avg}, is given by the expression $E[(n-2X)^2]$. By the properties of expected values (see, e.g., the book by Rothenberg in the references)

$$I_{\text{avg}} = E[(n-2X)^2] = E[n^2 - 4nX + 4X^2] = E[n^2] - 4nE[X] + 4E[X^2].$$
$$(81.2\text{a})$$

But for this problem $E[X] = n/2$ and $E[X^2] = (n + n^2)/4$, so that

$$I_{\text{avg}} = n^2 - 4n(n/2) + 4[(n + n^2)/4] = n. \qquad (81.2\text{b})$$

This means that on the average, given n sounds of unit amplitude with random phases confined to two opposite phases, their resultant intensity is always n. After lengthy calculations it is possible to show that the same result is valid if the phases are truly random (i.e., not confined to two values only). This result is only a mean intensity in a possible range from 0 to n^2; but if the changes are rapid, then the fluctuations from the mean are correspondingly small.

What, then, has been demonstrated here? Firstly, that the pitch of a composite note is approximately given by the average of those of its components, and, secondly, that the mean intensity is approximately the sum of the individual intensities. In practical terms this means that the pitch of the aeolian "whisper" of a pine tree, for example, is about the same as its average "leaf" or needle; furthermore, while the sound of an individual needle may be inaudible, that of the tree itself may be heard some distance away. One tree has many twigs or needles, but what about a forest of many trees? By the same arguments we can see that the vastly greater number of twigs/needles and trees will merge into the famous "murmur of the forest."

Q.82:   **How opaque is a wood or forest?** [color plate]

Suppose that you have walked some distance into a wood (perhaps the dog has slipped the leash again, or you just like to talk to the trees . . .). Looking

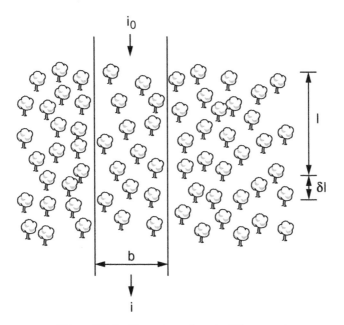

i_0

l

δl

b

i

Figure 82.1. Plan view of part of the wood

through a narrow strip of woodland, between the tree trunks you can see the light beyond, so how is the available light (incident from the edge of the wood) dependent on how far from the edge you are? We are not interested here in the light incident from above, but in the case of a thick leaf canopy at dusk or dawn, most of the available light does in fact come through from the perimeter of the wood.

At a distance l into the wood, consider a rectangular beam of light of width b and height h (figure 82.1). Suppose the trees are randomly distributed throughout the wood, with an average density of N trees per unit area, and on a level with our eyes they have an average diameter D. Furthermore, suppose that the amount of light per unit area normal to our line of sight (i.e., the light intensity) is initially i_0, and at a distance l into the wood it has been reduced to i (by the blocking effect of the trees). Then, going a little further into the wood (a distance δl) corresponds to a small change in light intensity δi, which is given by the area blocked off by trees in the sight "window." This is

$$\delta i = -\text{tree density} \times \text{area of ground} \times \text{tree thickness}$$
$$\times \text{"window" height}$$

or $\delta i = -Nb\delta lDh.$ \hfill (82.1)

Since at distance l into the wood, $i = bh$, it follows that

$$\frac{\delta i}{i} = -ND\delta l, \tag{82.2}$$

or, proceeding to the limit (which of course we can't strictly do, because trees are discrete objects) and using dummy variables, we have

$$\int_{i_0}^{i} \frac{d\xi}{\xi} = -\int_{0}^{l} ND d\eta, \tag{82.3}$$

or if N and D are constant,

$$i = i_0 e^{-NDl}. \tag{82.4}$$

Now we need to take a field trip to a wood to come up with some estimates. How about the Black Forest in southern Germany? Failing that, consider a wood of young fir trees, with (say) $N = 1$ per square yard and $D \approx 0.14$ yard (about 5 inches). Then we can construct the following table:

l yards	$\frac{i}{i_0} = e^{-0.14l}$
10	0.25
15	0.12
20	0.06
25	0.03

so 10 yards into the wood, the relative light intensity is 25%, and 25 yards in, it is only 3%. This is a striking increase of opacity. The reader can put in her own values for N and D from different locations, of course. But if we're going to traipse around the woods anywhere in Europe, we really need to use the metric system . . .

Another consequence of all this is that we can crudely solve a type of inverse problem. From a rough estimate of the fraction of the horizon that, from where we stand, is *not* intercepted by trees (this is proportional to i/i_0), from (82.4) we can use the formula

$$l = -\frac{1}{ND} \ln\left(\frac{i}{i_0}\right) \tag{82.5}$$

to approximate the depth of the wood (as far as we have traveled).

Question 83. A near-hemispherical tree "tumor."

Q.83:   Why do some trees have "tumors"?

Have you ever been walking in the woods, only to notice that some trees have spheroidal bumps and lumps on their trunks or branches? Some of these tumor-like protuberances can be quite large. They literally can be thought of as tumors, as the following description indicates:

> Tumors are growths of plants, animals or men in which the normal processes of control are, for some reason, ineffective, so that continued cell division results in massive disorganized development. When they are localized and not seriously inimical to the general functioning of the supporting body, they are called "benign." When their location or manner of development is such as to kill the supporting individual, they are called "malignant" and in man become "cancers." Tumors may arise from many causes, including parasites and infections of many kinds. The term "cancer," however, is generally limited to malignant tumors which arise from no clearly recognizable cause and in which some one or more cells of the host body have undergone a permanent change which renders them

unaffected by normal growth restraints so that the altered cell itself becomes a parasite or an infectious agent. Tumors exist in all classes of multicellular, organized living beings. In plants, essentially malignant tumors are known to be caused by hereditary factors, by certain bacteria, and by viruses.

So wrote Philip R. White, in a 1958 paper on an unusual type of tree tumor. Although the article is fifty years old, this general description remains remarkably accurate despite huge advances in the biochemical and genetic understanding of tumors and their causes. It is the general "uncontrolled cell proliferation" that I seek to model here, *in the early stages* of tumor growth. We do not necessarily need to restrict ourselves to trees, of course. The model is sufficiently simple that we can consider it to fit any context, animal or vegetable, at this point. Related to this, we shall briefly examine below a model that contains more biological input, and also has some historical relevance to the study of tumor growth in animals.

A TIME-DEPENDENT MODEL

If we are going to model the behavior of our generic "bole" or "tumor" over time (in an *extremely* simple fashion), then we need to set up an equation delineating how the tumor volume changes as a function of time t. Perhaps the simplest such model relates this to the rates at which the cells proliferate and die; in terms of a word equation, this is

$$\frac{d}{dt}\left(\begin{array}{c} \text{tumor} \\ \text{volume} \end{array}\right) =$$

$$\left(\begin{array}{c} \text{rate of volume increase due} \\ \text{to cell proliferation} \end{array}\right) - \left(\begin{array}{c} \text{rate of volume decrease} \\ \text{due to cell death} \end{array}\right) \quad (83.1)$$

First we identify the assumptions to be made in formulating this mathematical model:

(i) The tumor is spherically symmetric at all times, of radius $R(t)$ (often tree tumors tend to be hemispheroidal, but this is not a major problem here).

(ii) The rate at which new cells are formed is proportional to the rate at which the tumor receives nutrient.

(iii) The rate at which the tumor receives nutrient is proportional to its surface area. (Recall the snowball problem? This is a similar type of assumption to one in that model.) For the case of a tree tumor, this means that part of the its internal area in contact with the vascular system of the tree. In any case, in mathematical terms, we write this rate as $4\pi k_1 R^2$, k_1 being a constant of proportionality. Obviously, we could incorporate the 4π into this constant without loss of generality, but I like the number 4π . . .

(iv) The rate at which tumor cells die (through lack of nutrient) is proportional to the tumor volume, i.e., $4\pi k_2 R^3/3$, k_2 also being a constant of proportionality.

Under these circumstances, the word equation above becomes

$$\frac{d}{dt}\left(\frac{4}{3}\pi R^3\right) = 4\pi R^2 \frac{dR}{dt} \equiv 4\pi k_1 R^2 - \frac{4}{3}\pi k_2 R^3, \tag{83.2}$$

or

$$\frac{dR}{dt} = k_1 - \frac{1}{3}k_2 R = k(K - R), \tag{83.3}$$

where $k = k_2/3$ and $K = 3k_1/k_2$. (By the way, do you see *now* why we kept the 4π hanging around?) This equation is just crying out to be integrated, so we dutifully oblige: if R_0 is the tumor radius at some initial time $t = 0$, then

$$\int_{R_0}^{R} \frac{d\xi}{K - \xi} = \int_0^t k\,dt \Rightarrow \ln\left|\frac{K - R}{K - R_0}\right| = -kt, \tag{83.4}$$

from which we find that

$$|K - R(t)| = |K - R_0|e^{-kt}. \tag{83.5}$$

What are the implications of this result? Well, regardless of its initial size, the tumor eventually evolves to a limiting size K, since

$$\lim_{t \to \infty} R(t) = K. \tag{83.6}$$

More specifically we see that

$$R(t) = K + (R_0 - K)e^{-kt}, \tag{83.7}$$

as is readily verified. Therefore, if $R_0 < K$ (i.e., the initial tumor size is sufficiently *small*), then the tumor grows to a limiting size K; this is determined, of course, by specific values of the constants k_1 and k_2, if known. If, on the other hand $R_0 > K$ (the initial tumor size is sufficiently large), the tumor *shrinks* to the limiting size K. Note that increasing the value of k_1 (a measure of the rate at which tumor cells proliferate) and/or decreasing k_2 (a measure of the rate at which tumor cells die) will increase the limiting size K. Correspondingly, this size is decreased if k_1 is decreased and/or k_2 is increased.

Exercise: Derive the result (83.7) using the following short cut: let $\theta(t) = R(t) - K$.

Generally, of course, these "constants" are likely to be functions of the concentration of nutrient, which itself will depend on the location within the tumor. This spatial structure involves consideration of the diffusional aspects of the problem, but we will not pursue this here. However, it is worth noticing that this type of mathematical problem is in fact a very common one in many different contexts, and one such example is provided in appendix 3: the cooling of my cup of coffee!

In the national park

Q.84:   What shapes are river meanders?

I love looking at river meanders when I'm in a window seat on a plane. I've never been very successful at taking good photographs of them for at least three reasons: (i) I'm not the world's best photographer, (ii) even if I'm seated by a window, I'm often over a wing, and (iii) the amount of air between me and the ground causes much light to be scattered, so the scattered blue light reduces the contrast between the river and its surroundings. And if I'm walking in the vicinity of meanders and oxbow lakes, my perspective is very limited because I'm not one hundred feet tall. Regardless of my complaints about this, the phenomenon is a fascinating one, so let's start with some observational details, obtained from a study of more than fifty rivers by Luna Leopold and coworkers. They are of interest in connection with the apparent regular "sinuosity" of rivers the world over. From their study of more than 50 rivers, the following statements can be made:

  (i) No river, regardless of size, runs straight for more than 10 times its width.
 (ii) The radius of a bend is nearly always 2–3 times the width of the river at that point.
(iii) The wavelength (distance between analogous points of analogous bends) is 7–10 times the (average) width.

The conclusion is that despite considerable, even dramatic, variations in size and in bed conditions, rivers are strikingly similar in their characteristics. Furthermore, it transpires that meanders are not "accidents" of nature, but according to one theory, they define the form in which a river does the least work in turning (as in proceeding from a point A uphill to a point B downhill), which in turn defines the most probable form a river can take.

The name *meander* is from a winding stream in Turkey that was in ancient Greek times known as the Maiandros (today it is known as the Menderes). It

Question 84. River meanders viewed through an aircraft window.

might seem at first sight that local irregularities both within the river and at its boundaries are sufficient to divert the river from a straight course. Such irregularities, e.g., boulders, fallen trees, or harder rock through which the river flows, may indeed be sufficient to cause meanders, but they are not *necessary* conditions, being unable to account for the consistent geometry of meanders, for example, both in nature and in the laboratory. Meanders are observed in ocean currents (such as the Gulf Stream) and in water channels on the surface of a glacier. In rivers, the processes of continuous erosion, transportation, and deposition construct the ever-changing shape of rivers and their meanders.

Imagine a slight bend in a river. As water flows around it the "centrifugal force" will cause scouring of the outer, concave bank. (Note that this is a fictitious force, but one corresponding to a real phenomenon; it is the consequence of Newton's third law concerning action and reaction, and the existence of the *centripetal* force.) The water will then move downward at this bank (it has nowhere else to go) and then across the river bed toward the other bank. As it does this, friction with the bottom slows down the flow, allowing particles of silt and rock to be deposited there, though somewhat downstream due to the forward velocity of the river superimposed on this "crosswise" or "spanwise" flow. The actual fluid "particles" will execute helical paths as a result of this, and the scoured bank will be steeper than the gentler slopes on the other side of the river. This will result over time in the meander becoming more and more pronounced until to continue would result in the river moving uphill—a no-no for rivers! Thus, the turning process begins again as gravity continues to pull the river water to a level of lower potential energy. Such meanders will also "move" downstream over time. Apparently, the best conditions under which meanders arise is when the river traverses a gentle slope composed of fine-grained, easily erodable material that has sufficient cohesiveness to provide firm banks.

The researchers concluded that a given series of meanders tends to have a constant ratio between the "wavelength" of the curve (in its commonly understood sense) and its radius of curvature. If this ratio is fairly constant, the meanders appear, not surprisingly, to be rather regular also. Typical values range from about 5:1 for sine-type curves to 3:1 for more tightly looped meanders, the average over a sample of 50 typical meanders on different rivers and streams being about 4.7:1. The "tightness" of a bend (its *sinuosity σ*) is defined to be the ratio of the length of a channel in a given curve to its wavelength. This is rather variable, somewhere (for most meandering rivers) in the range 1.3:1 to 4:1.

From existing mathematical models in the literature the following curve is considered a good approximation to the shape of the center line for many meandering rivers:

$$\theta(l) = \theta_0 \cos\left(\frac{2\pi l}{L}\right), \tag{84.1}$$

where l is the length of the river from some initial point, L is a characteristic length associated with that particular river, and θ is the angle the central line makes with the down-river direction at each particular value of l. Note that (84.1) is *not* the empirical equation describing the "y"-location of the stream (if the x-axis is the center line of an imaginary straight stream); it describes how the change of direction of the meander relates to the distance along the curve.

Instead of getting into the rather advanced mathematics of meanders associated with angular dependence (84.1), we consider some less sophisticated models. We define the parameter α as the ratio of the arc length in a meander to its "wavelength." We can investigate this analytically easily enough for some simple meander shapes as follows. First we consider meanders to be

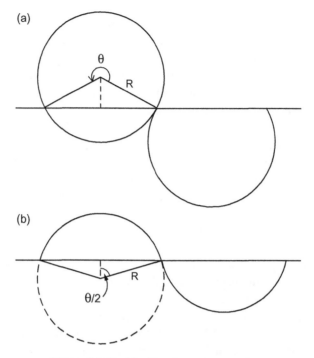

Figure 84.1(a,b). Circular arc meanders

composed of smoothly matched *circular* arcs, each of radius R and length l, where $0 < l < 2\pi R$ (see figure 84.1).

For this or any other configuration we need only to consider one such "loop" to determine α, and in this case

$$\alpha(\theta) = \frac{R\theta}{2R \sin \frac{\theta}{2}} = \frac{\theta}{2 \sin \frac{\theta}{2}}, \tag{84.2}$$

where θ is the angle (in the positive sense) shown in the figure, subtended by the two points of intersection of the meander with the x-axis. This definition works for both $\theta < \pi$ and $\pi < \theta < 2\pi$, of course, but in the latter regime an alternative possibility is to define the wavelength as the maximum width of the meander, i.e.,

$$\tilde{\alpha}(\theta) = \frac{R\theta}{2R} = \frac{\theta}{2}. \tag{84.3}$$

Obviously, $\tilde{\alpha}$ increases linearly, as opposed to the behavior of α, indicated in figure 84.2. They are equal when $\theta = \pi$ radians.

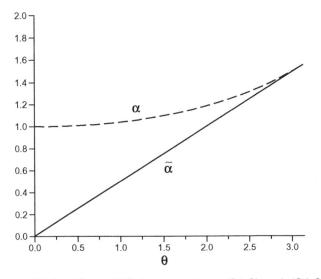

Figure 84.2. $\alpha(\theta)$ and $\tilde{\alpha}(\theta)$ from equations (84.2) and (84.3)

On the other hand, if we take a parabolic meander

$$y = (b/2)x^2, \qquad (84.4)$$

with origin at the vertex of the parabola, and $a > 0$, $b > 0$, then the *arc length* from the origin to the point $(\sqrt{2a/b}, a)$ is given by the integral

$$\int_0^{\sqrt{2a/b}} \sqrt{1 + \left(\frac{dy}{dx}\right)^2}\, dx = \int_0^{\sqrt{2a/b}} \sqrt{1 + b^2 x^2}\, dx$$

$$= \frac{\sqrt{2ba(1 + 2ba)} + \ln\left(\sqrt{2ba} + \sqrt{1 + 2ba}\right)}{2b}. \qquad (84.5)$$

Now the value of α is given by

$$\alpha = \frac{\sqrt{2ba(1 + 2ba)} + \ln\left(\sqrt{2ba} + \sqrt{1 + 2ba}\right)}{2\sqrt{2ab}}. \qquad (84.6)$$

Graphs of $\alpha(a; b)$ are shown in figure 84.3 for several values of b: $b = 1$ (solid); $b = 2$ (dashed) and $b = 0.5$ (dash-dotted); note also that α is symmetrical in a and b.

Although most of the mathematics for meanders based on equation (84.1) is beyond the scope of this book (see Adam (2006) for more details), some

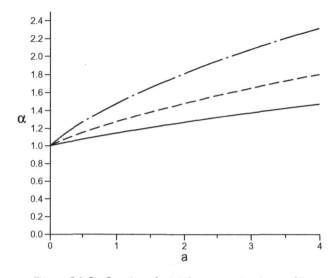

Figure 84.3. Graphs of $\alpha(a)$ for several values of b

insights can be gained from it. In one model, researchers studied an equation similar to (84.1), and provided an approximate empirical relation between the sinuosity σ and the initial angle θ_0 (in radians) as

$$\theta_0 = 2.2\sqrt{1 - \sigma^{-1}}. \tag{84.7}$$

(In degrees, the numerical factor of 2.2 becomes about 126°.) Inverting this expression,

$$\sigma = \left[1 - \left(\frac{\theta_0}{2.2}\right)^2\right]^{-1}. \tag{84.8}$$

In one model, the approximate expression for the *bend radius R* is found in terms of the wavelength L_x to be

$$R = \frac{L_x}{13}\frac{\sigma^{3/2}}{\sqrt{\sigma - 1}}, \tag{84.9}$$

from which the ratio

$$\Omega(\sigma) = \frac{L_x}{R} = 13\sqrt{\sigma - 1}\,\sigma^{-3/2}. \tag{84.10}$$

The graph of $\Omega(\sigma)$ increases from a value of zero on the interval $[1, \infty)$ to a maximum value of approximately 5 at $\sigma = 1.5$ and decreases toward zero thereafter. It is readily (but somewhat tediously) shown that, after a sequence of variable changes equivalent to $\sqrt{\sigma - 1} = \tan\theta$,

$$\int \Omega(\sigma)d\sigma = 26\ln|\sqrt{\sigma} + \sqrt{\sigma - 1}| - 26\sqrt{1 - \sigma^{-1}} + C, \qquad (84.11)$$

from which it follows that the average value of Ω on the interval $[1.1, 2]$ is

$$\frac{10}{9}\int_{1.1}^{2}\Omega(\sigma)d\sigma \approx 4.76. \qquad (84.12)$$

According to data used in one paper, for meanders in which sinuosities lie between 1.1 and 2.0, the average value of meander length to radius of curvature, Ω, is 4.7, showing good agreement between this model and actual measurements. Obviously, there are factors, such as variation in physical properties of the regions through which a river flows, that may modify the formation of meanders, but as anyone who has flown over land will have noticed, there are many fine examples of such quasi-periodic patterns to be seen from the air.

Q.85:   Why are mountain shadows triangular? [color plate]

When the Sun is low on the horizon, an observer near the summit of a mountain may be treated to a magnificent, if somewhat surprising sight: the mountain's shadow appears to be triangular, regardless of its actual profile! This effect is due to a combination of two things: perspective and the finite angular width of the Sun. Taking the latter first, as should be clear from figure 85.1, the Sun will produce both an umbral and a penumbral shadow from an opaque object such as a planet, moon, or mountain. Furthermore, since the Sun is larger than any of the objects, the umbral shadow is finite in length, and the effect of perspective makes the shadow appear to converge even faster to an observer on a mountain top. Probably one of the most common manifestations of this principle is when you notice your shadow stretching far ahead of you when the Sun is low on the horizon behind you; your tiny little "pinhead" is a long way from your feet, and some indication of "triangular convergence," as exhibited by the umbral shadow in figure 85.1, is apparent; of course, your limbs and head are still visible, though distorted through perspective.

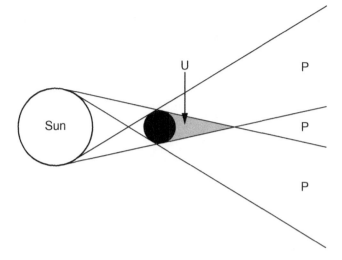

Figure 85.1. Umbral (U) and penumbral (P) shadows produced by the Sun.

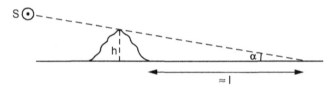

Figure 85.2. Basic mountain shadow geometry

So how long might such a shadow be? If we simply examine the length l of the projection of the shadow on the surface of a flat Earth, without regard to lateral convergence, we'll get a good estimate. For the Sun low on the horizon, say with altitude $\alpha \lesssim 5°$, and a mountain of height h km, we can easily see from figure 85.2 that

$$l \approx h \cot \alpha \approx \frac{h}{\alpha}, \tag{85.1}$$

where α is now in radians, and we have made the very reasonable assumption that the base half-width of the mountain can be neglected compared with l. Consider a mountain of height 4 km (or about 13,000 ft), with the Sun very low in the sky, i.e., $\alpha = 2°$ (or ≈ 0.035 radians); from (85.1), the shadow length is $l \approx h/\alpha \approx 110$ km. Doubling the height of the mountain for the same solar altitude would double the shadow length, and for a given mountain, at such small solar altitudes, halving the altitude would also double l. Of

course, at the summit of such a mountain, the distance to the horizon (neglecting atmospheric refraction) for an observer is provided by the approximate formula $d \approx \sqrt{1.5h}$ miles, h now being the height of the mountain in
feet (see questions 14 and 39). In this instance, then, $d \approx \sqrt{1.5 \times 13,000} \approx$
140 miles (≈ 220 km), so the shadow would still "hit the ground!" Thus, if
the Sun were on the horizon and the mountain had a cross section in the
shape of the Sydney Opera House or even Mickey Mouse, its shadow would
be a very long tunnel of unlit air converging almost to a point in the distance,
and the features of its shape would not be distinguishable to an observer on
the summit because of both its distance and the blurring effect of the Sun's
angular width.

With the aid of a telescope, these simple ideas can be used (in principle) to
estimate the height of mountains on the Moon, knowing the length of the
shadow on the lunar surface and the altitude of the Sun at that location, but
it's difficult to do on a night *walk!*

Q.86:   Why does Zion Arch appear circular? [color plate]

Among many other fascinating geological formations, a visitor to Zion National Park in Utah will notice many arches and alcoves in the rock. A natural
arch is created by natural geological forces causing rock to fall away, leaving
the arch structure standing. Rock, constantly under tensional stresses set up
by the pull of gravity, creates arch formations. Rock falls and slides are quite
common.

It is interesting to examine the effect of gravity versus the forces of adhesion in arches of various shapes. When an arch or alcove is formed by rock
shearing off under gravity, the adhesive forces have been broken by the weight
of the rock. We know that the lateral surface area of a cylinder is minimized,
for a given volume of rock, if the cylinder is circular. Equivalently, in two dimensions, the perimeter of a closed curve is minimized for a given area if that
curve is a circle, and any segment enclosed by a circular arc has the same
property. For a constant thickness of rock (assumed here) we may just use the
latter result.

In what follows, for simplicity, we shall ignore the shear stresses at the
"back" surface of an alcove, and consider only the orthogonal adhesive forces
associated with the bounding upper surface. This is clearly a major oversimplification, but if it encourages scientists and engineers to visit Zion National
Park, it will have been worth it! A more realistic model would take into account the varying vertical component of such forces along the boundary,
along with the neglected shear stresses.

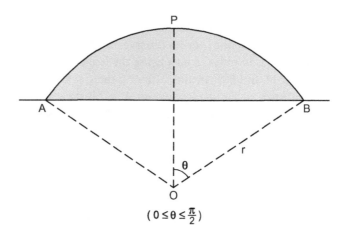

$$(0 \le \theta \le \tfrac{\pi}{2})$$

Figure 86.1. Zion Arch (circular) segment

From figure 86.1, where for obvious reasons $0 < \theta \le \pi/2$, the area of the sector $OAPBO$, by symmetry, is $r^2\theta$, whereas that of the segment $APBA$ is the area of the sector minus the area of triangle AOB: the area of this triangle (again by symmetry) is $r^2 \cos\theta \sin\theta$. Therefore, the area of the segment is $r^2 (\theta - \cos\theta \sin\theta)$ and the arc length APB is $2r\theta$; hence, the area-to-arc length (and also the weight-to-contact surface area) ratio is

$$\frac{r}{2}\left(1 - \frac{\cos\theta \sin\theta}{\theta}\right) = \frac{r}{2}\left(1 - \frac{\sin 2\theta}{2\theta}\right) \equiv \frac{r}{2}G(\theta). \qquad (86.1)$$

The function $G(\theta)$ is primarily of interest, since the $r/2$ term is just a scaling factor. Let's do a little curve sketching; although $G(\theta)$ is oscillatory, it is only so outside the range of physical interest (which was noted above to be $0 < \theta \le \pi/2$).

In fact,

$$G'(\theta) = \frac{\cos 2\theta}{2\theta^2}(\tan 2\theta - 2\theta) > 0 \text{ for } 0 < \theta < \theta_m, \qquad (86.2)$$

where $\theta_m \approx 2.25$ radians ($\approx 129°$) is the location of the first local maximum of G. Therefore, in the range of geological interest G increases monotonically away from its limiting value of zero as $\theta \to 0^+$.

Exercise: (i) verify the value of θ_m above, and (ii) show that at θ_m G is an *absolute* maximum for all values of θ.

Thus, when the angle θ is such that the weight of the rock segment exceeds the total upper surface sand-grain adhesive force, the segment shears off. Let us compare this with an open rectangular region, for example, of height L and width $(1+\alpha)L$, $\alpha \geq 1$. In this case, the area-to-perimeter ratio is $(1+\alpha)(3+\alpha)^{-1}L$. We know that a circular segment of the *same* area and thickness (and hence weight) will be the first to "drop" (compared with any other shape, in fact) because the contact surface area is less, but the weight-to-surface area ratios will be equal if

$$\frac{r}{2}G(\theta) = \frac{1+\alpha}{3+\alpha}L. \qquad (86.3)$$

For given values of any three of the parameters α, θ, r, and L the other can be determined; this might make an interesting exercise after a trip to Zion!

In the night sky

Q.87: **How are star magnitudes measured?**

When I was about fourteen or fifteen years old, my friend Miffy (a.k.a. Paul Smith) and I would amuse ourselves in our French class by doing calculations using some of the complicated looking formulas in Sir James Jeans' book *Astronomy and Cosmogony* [4]. While outdated (even then) as a textbook of astrophysics, it was fascinating both historically and scientifically, and I still have my copy of it. Since I was pretty good at French, I had done my home-work, and as long as we were quietly working on the back row, our teacher probably had no idea that we were budding astrophysicists . . . besides, the French books we were using were old and dog-eared, and my lovely Dover edition of Jeans' classic 1928 work was crisp and new. It was very exciting to skim through its pages and encounter very impressive formulae; at the time I had no idea what most of them meant, but I determined that one day I would (and one day I will.)

The first of the formulas in chapter two of the book (*The Light from the Stars*) enabled one to calculate the differences in brightnesses (*apparent magnitudes*) two stars would have in terms of the light received from those stars (assuming that they are both equally distant from us). The fainter a star appears, the larger is its magnitude. Let us examine this concept in terms of a standard logarithmic model of the magnitude scale. As noted in question 41, the observed brightness of stars is expressed in terms of their *apparent magnitudes m* on a numerical scale that increases as the brightness decreases:

$$m = 6 - 2.512 \log_{10}\left(\frac{L}{L_0}\right), \tag{87.1}$$

where L is the light flux (luminosity or brightness) of the star (or planet) and L_0 is the brightness of the faintest star visible to the (average human) naked eye. Thus, when $L = L_0$, $m = 6$, and, therefore, since

$$L = L_0(10^{(6 - m)/2.512}), \qquad\qquad (87.2)$$

it follows that a star of magnitude $m = 1$ is ≈ 100 times brighter to the eye than a star of magnitude $m = 6$. In practice, any star for which $m \leq 1$ is referred as being of first magnitude, and of second magnitude if $1 < m \leq 2$, etc. Thus, Sirius, the Dog Star, with $m = -1.6$ is the brightest star in our sky apart from the Sun, Arcturus in the constellation of Boötes is first magnitude with $m = 0.2$, and the polestar with $m = 1.9$ is of second magnitude (values differ slightly depending on which astronomical almanac one consults!) The planets, of course vary in magnitude as the distance between them and the Earth changes.

So from where does this quantitative concept of star magnitudes come? In the second century BC Hipparchus introduced the concept of magnitude as a measure of stellar brightness, and later Ptolemy (in the second century AD), in the *Almagest*, divided the stars into six groups each of a different magnitude. The 20 brightest stars were deemed to be of first magnitude, and the faintest stars were designated as sixth magnitude. What Ptolemy regarded as five equal differences in brightness (1st to 6th magnitude) correspond to differences in the *logarithms* of light received (according to a standard physiological stimulus-response "law" (*Fechner's* law). (Actually, such responses are rather better described by power laws as more recent research suggests, but the logarithmic description is quite adequate for our purposes.)

In 1830 Sir John Herschel remarked that a star of magnitude 1 was about 100 times as bright as a star of magnitude 6, so that Ptolemy's five steps corresponded to a difference of two in the logarithms of the light received, and each unit magnitude difference corresponded to a brightness difference of $(100)^{1/5} \approx 2.512$. Later, in 1856, Norman Pogson made this approximate relationship to be an exact one, thus redefining the notion of magnitude. Therefore, a generalization of equation (87.1) above becomes, for any two stars of magnitudes m_1 and m_2, respectively,

$$m_1 - m_2 = -2.512 \log_{10}\left(\frac{L_1}{L_2}\right). \qquad\qquad (87.3)$$

The *absolute magnitude M* of a star is the apparent magnitude it would have if it were placed at a distance from us of 10 parsecs (about 32.6 light years). This particular distance is based on the concept of astronomical parallax, and the interested reader can pursue this further! One parsec is the distance of a star that would have a parallax of one second of arc subtended at the Earth over a distance equal to the average Earth–Sun distance. The point here is that we can use the inverse-square law of light intensity to relate M to the star's distance r (in parsecs). Making the obvious changes in equation (87.1) for a star of brightness (or luminosity) $L(r)$ we have that

$$m - M = -2.512 \log_{10} \left[\frac{L(10)}{L(r)} \right], \tag{87.4}$$

and from the inverse-square law $L(r) \propto r^{-2}$ it follows that

$$\frac{L(10)}{L(r)} = \left(\frac{r}{10} \right)^2. \tag{87.5}$$

Therefore,

$$m - M = 5.024 \log_{10} \left(\frac{r}{10} \right). \tag{87.6}$$

For the sun, with a whopping value for $m (= -26.7)$, it transpires that $M = +4.9$, so good ol' Sol would be barely visible to the naked eye at the standard distance of 10 parsecs.

Q.88:   How can I stargaze with a flashlight?

Have you ever used a flashlight beam as a pointer to identify particular stars, star groups, or planets to interested bystanders? Perhaps there were none. No matter; you may have noticed that even if the sky is clear, the beam seems to come to an end very abruptly in a particular direction. But why do we see the beam in the first place? The explanation is the same as that for the visibility of sunbeams: dust particles and water droplets in the beam scatter the light, some of which, of course, enters the eyes of the observer. In figure 88.1, the observer—you—is at the point O and the beam of light starts at L and extends (while diverging somewhat) to C and beyond. From points A, B, C, etc., in the forward direction, light is scattered toward the observer, but no matter how much the beam extends, the observer, displaced from it, will never see it extend beyond the direction OD (parallel to the axis of symmetry of the beam). Of course, the closer the direction of sight to that direction, the wider the beam is, and the more scattering particles there are, and this will to some extent counteract the fact that the beam is more remote from O in that direction.

Another comment is in order. If you look in the direction OA' and compare the intensity of scattered light with that when you look in the direction

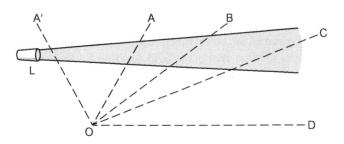

Figure 88.1. Geometry of the flashlight beam

OB (about 90° from *OA'*), you will notice that the former is considerably greater: more light is scattered in the forward direction than in the backward one. The explanation is that the particles scattering the light are sufficiently large that they scatter asymmetrically: much more light is scattered in the forward direction. This is essentially why the color of the blue sky can change as a result of changing proportions of dust, ash, salt particles, and water droplets in the atmosphere, and it also accounts for some of the differences between a blue sky in southern Europe, or the tropics, and a blue sky in northern England. Basically, small particles (of size $\approx 10^{-8}$ m) scatter blue and violet light most, and with almost equal intensity in all directions, whereas large particles (of size $\approx 10^{-6}$ m) scatter all colors more or less equally, but mostly at small scattering angles (i.e., in the forward direction).

Q.89:   How can you model a star?

In one of the scenes in the movie *The Lion King*, Timon the meercat and Pumbaa the warthog are relaxing on the ground, looking up at the stars. Timon explains that the stars are, in fact, fireflies. "Fireflies that, er, got stuck up in that big bluish black thing." Apparently awed by this information, Pumbaa replies, "Oh, gee, I always thought they were balls of gas burning billions of miles away." Clearly, Pumbaa had a much better grasp of the stellar heavens than did Timon. But at least they were both looking up and enjoying the night sky. When my children were much younger our family would often spend a week in the summer staying in a cabin at a state park. The night skies were usually cloud free, and in the darkness the Milky Way seemed like a bright path gently meandering though a diamond-studded velvet carpet. I became fascinated by the stars as a boy of twelve, and acquired a beautiful (but dented) brass three-inch refractor two years later. As you may recall from an earlier question, I spent many a pleasant French class playing with star-related formulae . . .

Let's start with a basic question. There must be something preventing the Sun or most other stars from collapsing under their own weight, right? Clearly, a star (or a planet for that matter) is held together by the force of gravitation, but without resistance, this would cause the star to collapse in a relatively short time (for the Sun, this is about 27 minutes; see question 90). The resistance is provided by the thermal *pressure* of the "star stuff" or stellar material (plasma), and these two physical effects must be in approximate balance for stars to remain stable over long periods of time. One fundamental assumption made here is that the stellar properties do not change with time; certainly not over the time you take to read this chapter! We will not get heavily into the physics here, but mention just enough to enable us to develop some simple models of nonrotating stars (and though all stars probably rotate, in many cases it is not necessarily significant: a little more will be implied about this in question 90 for the final exercise).

Imagine moving out a distance r from the center of a spherically symmetric nonrotating star, where $r \leq R$, R being the radius of the star (assumed well defined here). If $M_r(r)$ is the mass contained within a sphere of radius r, then the mass δM contained within a thin, spherical shell of thickness $\delta r \ll r$ is defined by

$$\delta M_r \equiv M_r(r + \delta r) - M_r(r) \approx 4\pi r^2 \rho(r) \delta r, \tag{89.1}$$

where $\rho(r)$ is the density of the stellar material. Provided that the limit exists,

$$\frac{dM_r}{dr} = \lim_{\delta r \to 0} \frac{\delta M_r}{\delta r} = 4\pi r^2 \rho(r). \tag{89.2}$$

This is sometimes called the equation of mass conservation. Obviously, $M_r(r)$ increases with r. An integral version of this equation is

$$M_r = \int_0^r 4\pi \xi^2 \rho(\xi)\, d\xi. \tag{89.3}$$

Next, we will consider the equation describing the balance between the pressure gradient within the star and the gravitational force per unit volume. Since the pressure acts radially *outward* (from below) on a spherical shell and *inward* (from above), then the pressure difference across the shell (or difference in force per unit area) is $\delta P = P(r) - P(r + \delta r)$. On the other hand, the inward gravitational force between a shell of mass δm and the mass *interior* to the shell is provided by the expression $-GM_r\, \delta m / r^2$, G being the constant of universal gravitation. Note that, by symmetry, the net gravitational force on the shell from material *outside* the shell is zero (though we do not prove that here). Since for each such element of mass (per unit area) $\delta m = \rho(r)\,\delta r$, we may use this to express the equilibrium configuration of a star by equating the forces on the shell due to pressure difference and gravity. Therefore,

$$\delta P = -\frac{GM_r}{r^2}\rho(r)\delta r, \qquad (89.4)$$

from which

$$\frac{dP}{dr} = \lim_{\delta r \to 0} \frac{\delta P}{\delta r} = -\frac{GM_r}{r^2}\rho(r). \qquad (89.5)$$

This is the equation of hydrostatic equilibrium. By definition it is a statement that all the forces on every "piece" of the star are in balance, i.e., the net force is zero. Since the pressure gradient is negative, pressure decreases outward from the central region of our star, as one would expect.

Next, we move on to energy conservation (a very good idea). All life on Earth owes its existence to solar energy. This energy drives the whole shebang, so to speak, from microbes to hurricanes, and obviously that energy must come from somewhere. That "somewhere" is the ongoing nuclear fusion reactions taking place in the heart of the Sun and other stars. Again, we consider our old friend, the spherical shell, and derive an equation relating the rates of energy release and transport. This time we denote by $L(r)$ the radial energy flow rate across the surface of a sphere of radius r (in appropriate units, e.g., watts). Suppose further that energy is released per unit mass at a rate $\varepsilon(r)$ at radius r; then the difference in energy flow across our thin spherical shell must be balanced by the energy produced or released within it. Then

$$L(r + \delta r) - L(r) = 4\pi r^2 \rho(r)\,\varepsilon(r)\,\delta r, \qquad (89.6)$$

and so

$$\frac{dL}{dr} = \lim_{\delta r \to 0} \frac{\delta L}{\delta r} = 4\pi r^2 \rho(r)\,\varepsilon(r). \qquad (89.7)$$

This equation tells us that the energy flow rate increases outward. But what about energy transport—how does it take place? Since the Sun is hotter at the center than at the surface (photosphere), heat energy flows outward. (It must be pointed out, though, that the solar corona, while very tenuous compared with the photosphere, is much hotter, being of the order of at least a million degrees Celsius [and more so above sunspots], as opposed to 5700 degrees Celsius at the photosphere.) But that's another story.

Energy can be transferred from one place to another by three fundamental mechanisms: conduction, convection, and radiation. Convection is more of a large-scale process, and is difficult to quantify accurately for our purposes, whereas the other two mechanisms are small-scale in nature, involving exchange of energy between particles (via collisions). In conduction, it is electrons that transport energy, but in radiation, it is by photons that the job is done, and in most cases this is the dominant mechanism. For each of these two processes, the energy flux (or energy per square meter per second) $F(r)$ can be written in terms of the temperature $T(r)$ in the form

$$F(r) = -\kappa \frac{dT}{dr}, \tag{89.8}$$

where κ is a coefficient of conductivity. In general, it is dependent on several stellar properties. The negative sign in this equation ensures that heat is transferred from regions of higher temperature to regions of lower temperature, as it should. A quantity encountered above, the radial energy flow rate $L(r)$, is related to the flux via the expression

$$L(r) = 4\pi r^2 F(r), \tag{89.9}$$

so that

$$\frac{dT}{dr} = -\frac{F(r)}{4\pi r^2 \kappa}. \tag{89.10}$$

A Simple Stellar Model

Many mathematical models involving varying physical or biological quantities incorporate linear behavior as a simple deviation from constancy. However, even then, the models may become quite complex mathematically. In this section, we suppose that the density of the star varies linearly from the center of the star to the "surface"; of course, this is a *very* unrealistic assumption, but nevertheless it will give us some valuable qualitative insights into the properties of real stars. So if ρ_c is the central density for our star of radius R, then

$$\rho(r) = \rho_c \left(1 - \frac{r}{R}\right). \tag{89.11}$$

Now for some calculations using the equation of hydrostatic equilibrium (89.5). First, though, we need to obtain an expression for M_r in terms of the density, so from equation (89.2)

$$\frac{dM_r}{dr} = 4\pi r^2 \rho_c \left(1 - \frac{r}{R}\right), \tag{89.12}$$

which upon integration becomes

$$M_r = \frac{4\pi\rho_c}{3} r^3 - \frac{\pi\rho_c}{R} r^4, \tag{89.13}$$

the constant of integration being zero. The total mass M of the star follows directly:

$$M \equiv M_r(R) = \frac{\pi\rho_c}{3} R^3, \tag{89.14}$$

and this provides us with an expression for ρ_c in terms of the mass of the star. Now from equation (89.5) we are able to express the pressure gradient as

$$\frac{dP}{dr} = -\pi G\rho_c^2 \left(\frac{4}{3}r - \frac{7r^2}{3R} + \frac{r^3}{R^2}\right). \tag{89.15}$$

Upon integration, the pressure distribution within our "linear" star is found to be, in terms of the central pressure P_c,

$$P(r) = P_c - \pi G\rho_c^2 r^2 \left(\frac{2}{3} - \frac{7r}{9R} + \frac{r^2}{4R^2}\right). \tag{89.16}$$

It follows that the pressure at the surface of the star is

$$P(R) = P_c - \frac{5}{36}\pi G\rho_c^2 R^2, \tag{89.17}$$

but this is by definition zero, and hence the central pressure is

$$P_c = \frac{5}{36}\pi G\rho_c^2 R^2. \tag{89.18}$$

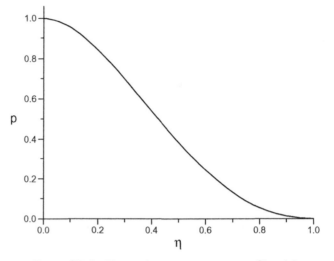

Figure 89.1. Dimensionless pressure profile $p(\eta)$

Thus, the central pressure is proportional to both the square of the radius and the square of the central density according to this model. Using (89.14), this may also be expressed in terms of the stellar mass as

$$P_c = \frac{5}{4\pi} \frac{GM^2}{R^4}. \tag{89.19}$$

On combining this with equation (89.16) we can express the pressure distribution rather nicely in terms of the relative (and hence dimensionless) radius $\eta = r/R$ and pressure $p(\eta) = P/P_c$:

$$p(\eta) = 1 - \frac{24}{5}\eta^2 + \frac{28}{5}\eta^3 - \frac{9}{5}\eta^4, \tag{89.20}$$

depicted in figure 89.1.

We can also obtain an expression for the temperature distribution if we ignore something called radiation pressure. This cannot be ignored in reality, but it would take us too far into the physics of stellar interiors to incorporate this, and it would, in that immortal phrase, "lie outside the scope of this book!" Remember, we are playing with models here in order to learn to model phenomena mathematically. We shall therefore invoke the so-called *perfect gas law* for which the temperature

$$T(r) = k\frac{P(r)}{\rho(r)}, \tag{89.21}$$

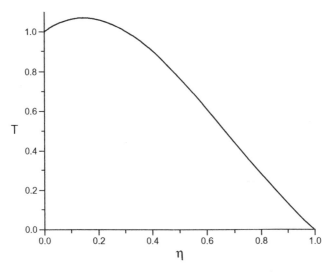

Figure 89.2. Dimensionless temperature profile $T(\eta)$

the constant k being of no concern to us here. From this it follows that the temperature at the center of the star is $T_c = kP_c/\rho_c$ and so, in terms of η,

$$\frac{T}{T_c} = \left(\frac{P}{P_c}\right)\left(\frac{\rho_c}{\rho}\right) = 1 + \eta - \frac{19}{5}\eta^2 + \frac{9}{5}\eta^3, \qquad (89.22)$$

after a modicum of long division! Again, this quantity is depicted in figure 89.2. Interestingly, this indicates that the maximum temperature occurs, not at the center of the star, but at a relative radius of

$$\eta = \frac{19 - \sqrt{226}}{27} \approx 0.147. \qquad (89.23)$$

Exercise: Prove this result.

We cannot use this, however, to draw any significant conclusions about the temperature structure of real stars because of the neglect of radiation pressure in this model.

We now return to a more "general" star and draw some surprisingly strong conclusions about its minimum central pressure, without any knowledge about the composition of the star; all that is required is knowledge of its mass and radius. Similar conclusions may be drawn about the minimum central temperature, but this would involve a discussion of something called the *virial theorem*, and take us too far afield. From (89.2) and (89.5), using the chain rule, we find that

$$\frac{dP}{dM} = \left(\frac{dP}{dr}\right)\left(\frac{dM}{dr}\right)^{-1} = -\frac{GM}{4\pi r^4}. \tag{89.24}$$

Integrating this equation with respect to M between its center (subscript c) and surface (subscript s) produces

$$-\int_0^{M_s} \frac{dP}{dM}\, dM = P_c - P_s = \int_0^{M_s} \frac{GM}{4\pi r^4}\, dM, \tag{89.25}$$

where now P_s is not necessarily zero, and M_s is, of course, the total mass of the star. The next step is to use an inequality based on the fact that $r < r_s\,(= R)$ at all points within the star (this is not rocket science), so $r^{-4} > r_s^{-4}$. Therefore,

$$\int_0^{M_s} \frac{GM}{4\pi r^4}\, dM > \int_0^{M_s} \frac{GM}{4\pi r_s^4}\, dM = \frac{GM_s^2}{8\pi r_s^4}. \tag{89.26}$$

Upon combining this with (89.25) we immediately have the result

$$P_c > P_s + \frac{GM_s^2}{8\pi r_s^4} \geq \frac{GM_s^2}{8\pi r_s^4}. \tag{89.27}$$

This applies to any star, of course, and in particular to the Sun, for which we know that $M_s \approx 2\times10^{30}$ kg and $r_s \approx 7\times10^8$ m. We shall denote these by $M_\odot$ and $r_\odot$, respectively. Additionally, the gravitational constant $G \approx 6.7\times10^{-11}$ N m^2/kg^2. Therefore, if we denote P_c for the Sun by $P_\odot$, then

$$P_\odot \geq \frac{(6.7\times10^{-11})\times(2\times10^{30})^2}{8\pi\times(7\times10^8)^4}\, \text{N/m}^2 \approx 4.4\times10^{13}\,\text{N/m}^2, \tag{89.28}$$

or equivalently,

$$P_\odot \geq 4.4\times10^8 \text{ atmospheres}, \tag{89.29}$$

since one atmosphere is approximately 10^5 N/m^2 (or 10^5 Pa). The inequality (89.27) can be rewritten for stars other than the Sun as

$$P_c \geq \frac{GM_S^2}{8\pi r_S^4} = \frac{GM_\odot^2}{8\pi r_\odot^4} \left(\frac{M_s}{M_\odot}\right)^2 \left(\frac{r_\odot}{r_s}\right)^4 = P_\odot \left(\frac{M_s}{M_\odot}\right)^2 \left(\frac{r_\odot}{r_s}\right)^4$$

$$= 4.4 \times 10^{13} \left(\frac{M_s}{M_\odot}\right)^2 \left(\frac{r_\odot}{r_s}\right)^4 \text{N/m}^2. \tag{89.30}$$

Recall that the question asked, *How can you model a star?* Starting from basic principles and using some simple ideas, we have been able to obtain some elementary information about the structure of a "toy" star. This last result, however, is surprisingly general. It expresses the minimum central pressure of any star, provided we know only the ratios of its mass and radius to those of the Sun. And, of course, it provides a lower bound to the central pressure of the Sun immediately, though it is too low by a factor of several hundred. However, given the crudity of the model (and it *is* a lower bound!), it's not bad at all.

Q.90: How long would it take the Sun to collapse?

From the equation of hydrostatic equilibrium (89.5), describing the balance between the pressure gradient and gravity in a star in equilibrium, we have

$$\frac{dP}{dr} = -\frac{GM_r}{r^2}\rho(r). \tag{90.1}$$

Now suppose first of all that these forces are not in balance:

$$-\frac{dP}{dr} - \frac{GM_r}{r^2}\rho(r) \neq 0, \tag{90.2}$$

By Newton's second law of motion (i.e., force equals rate of change of momentum, or, here, just $F = ma$), this imbalance at a given radial shell r can be written in terms of acceleration, i.e.,

$$\rho(r)\frac{d^2r}{dt^2} = -\frac{dP}{dr} - \frac{GM_r}{r^2}\rho(r). \tag{90.3}$$

If there were no pressure gradient to resist the gravitational force, the Sun (or other star) would collapse; this corresponds to

$$\frac{d^2r}{dt^2} = -\frac{GM_r}{r^2}.$$ (90.4)

For any given density profile $\rho(r)$ this equation can be solved in principle, but that is not necessary for our purposes. Since the acceleration is negative (the gravitational force is radially inward),

$$\frac{d^2r}{dt^2} = -\left|\frac{d^2r}{dt^2}\right|,$$

and dimensionally, from (90.4),

$$\frac{R}{T^2} \sim \frac{GM}{R^2},$$ (90.5)

where R and T are characteristic length- and timescales for the star in this collapse scenario, and M is the mass of the star. Therefore,

$$T \sim \left(\frac{R^3}{GM}\right)^{1/2}.$$ (90.6)

In astrophysical jargon, T is known as the dynamical timescale; it is a measure of the time it would take for the "surface" of the star to "freely fall" to $r = 0$. For the Sun we have used the values $M \approx 2 \times 10^{30}$ kg and $R (= r_s$ in question 89) $\approx 7 \times 10^8$ m, and if we substitute these values, and the value of G into (90.6) we find that

$$T \sim \left(\frac{[7 \times 10^8]^3}{6.7 \times 10^{-11} \times 2 \times 10^{30}}\right)^{1/2} \text{ s} \approx (2.56 \times 10^6)^{1/2}\text{s} \approx 1600\,\text{s},$$

so T is approximately 27 minutes! That is a pretty fast collapse time, especially when compared with the "nuclear burning timescale" for stars like the Sun, which is about 10^{10} years!

We can take these arguments a little further. If we define an average density $\bar{\rho}$ such that (using equation (89.3))

$$M = M_R = \int_0^R 4\pi\xi^2\rho(\xi)\,d\xi \equiv \frac{4}{3}\pi R^3\bar{\rho},$$ (90.7)

i.e.,

$$\bar{\rho} = \frac{3}{R^3} \int_0^R \xi^2 \rho(\xi)\, d\xi, \tag{90.8}$$

then from (90.6) and (90.7) we find that the free-fall timescale is

$$T \sim \left(\frac{3}{4\pi G\bar{\rho}}\right)^{1/2}. \tag{90.9}$$

We conclude with several interesting exercises, based on a web article by Shude Mao (see references).

Exercise: Estimate the free-fall time for a red giant with $R \approx 100R_{\odot}$ and $M \approx 1M_{\odot}$.

Exercise: Instead of the pressure gradient within a star disappearing, consider the opposite extreme: its gravity disappears! The star will "explode," rather than "implode!" Estimate the timescale for this to occur in terms of R, ρ, and P.

Exercise: Show that, for a uniform sphere with density $\bar{\rho}$, the free-fall time can be derived exactly, and is given by $T_e = (3\pi/32G\bar{\rho})^{1/2}$. Compare this with the result (90.9), and show that

$$\frac{T}{T_e} = \frac{2\sqrt{2}}{\pi}. \tag{90.10}$$

Comment on this result. Our final exercise has some bearing on the influence (or not) of rotation in departure from spherical symmetry in the Sun.

Exercise: The rotation period of the Sun is about 27 days. Estimate the ratio of the centripetal acceleration and the gravitational acceleration at the solar equator, and comment on the significance of this result.

Q.91:   **What are those small rings around the Moon?** [color plate]

A beautiful phenomenon sometimes visible on a clear night with a full Moon (or close to full) is a lunar corona. It consists of several softly colored rings

often apparent when only a layer of thin cloud covers the Moon without obscuring it. A corona can also appear around the Sun (or, in principle, around
any sufficiently bright extended source of light), but this should never be
looked for directly, because of the danger of permanent damage to eyesight.
Also, this corona is not to be confused with the *solar corona*, the tenuous outer
atmosphere of the Sun visible during a total eclipse of the Sun, mentioned in
question 89.

In the article by Cowley et al., the colors are described as follows: "At the
center is a very bright *aureole*, almost white, and fringed with yellow and red.
Sometimes that is all there is to be seen, but the better coronae have one or
more successively fainter and gently coloured rings surrounding the aureole.
The first is bluish on the inside, grading through greens and yellows to red
outermost . . . All the colours are subtle mixtures rather than the more direct
hues of the rainbow."

What causes a corona? The culprit, so to speak, is diffraction: a consequence
of light being obstructed by an obstacle comparable in size to the wavelength
of the light (see questions 47 and 48 for earlier mention of diffraction). In
this case, the obstacles are myriads of cloud droplets, ranging in size from
diameters of 1–100 microns, with mean diameters of between 10 and 15 microns. While this is an order of magnitude larger than the wavelengths (λ) of
visible light (0.45 to 0.70 microns) it is still sufficient for the effects of diffraction to be apparent, much like the bending of sound around pillars in an
auditorium. The much larger raindrops do not give rise to observable diffraction phenomena, though, of course, they do "scatter" light to produce the
rainbow.

The intensity $I(\theta)$ of diffracted light as a function of scattering angle θ
for an obscuring disc of radius R is expressed in terms of a very interesting
function (called a Bessel function; see appendix 1) that is denoted, for historical reasons, by $J_1(x)$, where in this context

$$x = \frac{2\pi R \sin\theta}{\lambda}, \text{ or } \theta = \arcsin\left(\frac{x\lambda}{2\pi R}\right). \tag{91.1}$$

Clearly, x is proportional for a given value of θ to the ratio of the circumference of the obscuring disc to the wavelength of light impinging upon it.
In fact,

$$I(\theta) = I(0)\left(\frac{2J_1(x)}{x}\right)^2, \tag{91.2}$$

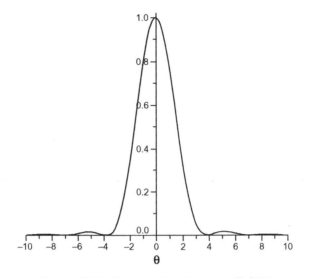

Figure 91.1. The relative intensity $I(\theta)/I(0)$

$I(0)$ being the intensity of light in the "forward" direction ($\theta=0$), and it is proportional to the area of the disc. The relative intensity $I(\theta)/I(0)$ is plotted in figure 91.1 as a function of x.

As can be seen, this quantity is oscillatory, but with a large central amplitude; successive maxima are very much smaller, corresponding to a rapid decrease of ring intensity as we look radially outward from the aureole. In fact, for $x \neq 0$ the first and second maxima are respectively only 1.75% and 0.42% of the central maximum at $x=0$. These occur respectively at $x=5.14$ and $x=8.42$, from which we can determine that the angles θ_1 and θ_2 for the first and second maxima are given by

$$\theta_1 = \arcsin\left(\frac{x\lambda}{2\pi R}\right) = \arcsin\left(\frac{5.14}{2\pi}\frac{\lambda}{R}\right) = \arcsin\left(0.82\frac{\lambda}{R}\right), \text{ and} \qquad (91.3)$$

$$\theta_2 = \arcsin\left(\frac{8.42}{2\pi}\frac{\lambda}{R}\right) = \arcsin\left(1.34\frac{\lambda}{R}\right). \qquad (91.4)$$

Of course, it must be remembered that just as a rainbow is the result of scattering of sunlight by myriads of raindrops, so the lunar corona is a result of diffraction by a great many cloud droplets. When these droplets are all more or less the same size these diffraction effects reinforce each other and the resulting coronae are usually bright and well delineated. This degree of droplet uniformity is present when the cloud is rather "new," i.e., the droplets have the same "history"; small environmental differences affecting individual droplets have not had time to accumulate. According to Cowley et al., such conditions are found most often in altocumulus, cirrocumulus, and lenticular clouds.

On the other hand, if there is a wide range of droplet size in the cloud, the coronae tend to be washed out because the diffraction patterns overlap, and only the aureole is easily seen. Another variation occurs if the size distribution is narrow, but differs from region to region in the cloud; this may result in coronal arcs of different radii and, hence, noncircular, indeed, irregular arcs may be seen. If the patches are sufficiently random, the corona may result in what is referred to as cloud iridescence; this is frequently seen during the day in clouds in the vicinity of the Sun (see question 48).

Q.92: How can you model an eclipse of the Sun?

As a child, interested in astronomy from my preteen years, I always dreamed of witnessing a total eclipse of the Sun. The one I looked forward to was on August 11, 1999, some thirty-eight years ahead of me at that time. If the skies were clear, it would be visible from Cornwall and the southern part of Devon, being partial over the remainder of the United Kingdom. The previous total eclipse had been on June 30, 1954 (when I was too young to really appreciate such things), totality being visible only from the Scottish island of Unst, in the Shetland Islands. Again, not surprisingly, a partial eclipse was visible from the rest of the United Kingdom.

But back to less prehistoric times; in 1984 my family and I moved perma-nently (thus far at least) to the United States. There was nothing, in principle, that would have stopped me from flying over to witness this long-sought event, but I did not go; oh, how youthful dreams can fall prey to the busy-ness and apathy of middle age! As it happened, the eclipse was a great disappoint-ment for most people; that part of the sky was almost completely cloudy for most observers, though a partial eclipse was seen briefly from Alderney in the Channel Islands.

The next total solar eclipses (at the time of this writing) through the end of 2020 will take place as follows (whether anyone can see them or not!):

  (i) August 1, 2008 (in parts of northern Canada, Greenland, Siberia, Mongolia, and China)
  (ii) July 22, 2009 (in parts of India, China, and the Pacific Ocean)
  (iii) July 11, 2010 (in parts of South America and Tahiti)

(iv) November 13, 2012 (in parts of Australia, New Zealand, South America, and the southern Pacific)
(v) March 20, 2015 (Atlantic Ocean, Norway, and the North Pole)
(vi) March 9, 2016 (in parts of southern Asia and the Pacific Ocean)
(vii) August 21, 2017 (in the United States from Oregon to South Carolina)
(viii) July 2, 2019 (in parts of South America)
(ix) December 14, 2020 (also in parts of South America)

It would seem that the best bet for those living in the United States who are not able to travel to other locations would be the one in 2017, though the one visible from Tahiti in 2010 certainly appeals to me!

A recent scientific article (by Mollmann and Vollmer) concerns the changes in *illuminance* during a partial or total eclipse. Illuminance (which will be denoted by L) is a rather technical photometric term; it is sufficient for our purposes to note that it gives a measure of illumination as perceived by a "typical" human eye. Its units of measurement are usually given in terms of either *lux* (lx) or lumens per square meter. To illustrate the typical range of illuminances, that for the full moon is about 0.25 lx; street lights tend to vary between 1 and 16 lx (in Germany, at least—these data are taken from the paper by Mollmann and Vollmer); living room lights are about 120 lx; good working room conditions are about 1000 lx, and, depending on the solar elevation and cloud thickness, an overcast sky is of order 10,000 lx, while a cloudless sky can be at least 70,000 lx, again, depending on solar elevation.

The mathematical model proposed assumes first of all that during an eclipse, the illuminance is directly proportional to the unobscured area of the Sun's disk (there is a small contribution from skylight (scattered sunlight); it is a relatively small contribution that will be neglected in this simple model. Other assumptions are the following:

(i) The Sun and Moon are treated as circular disks (of *angular* radii r and R, respectively), thus neglecting, among other features, the presence of lunar mountains, which can give rise to the spectacular "diamond ring effect" a second or two before totality. Note that a necessary condition for an eclipse to be total is that $R \geq r$. Of course, this is not a sufficient condition, and it is important to note that even when the Sun, Moon, and Earth are collinear, r can lie in the interval $0.90R \leq r \leq 1.06R$, because of variable distance of the Moon from the Earth (and assuming r is constant, for simplicity).

(ii) The paths of the Sun and Moon as viewed from the Earth are assumed to be straight lines for the duration of the eclipse.

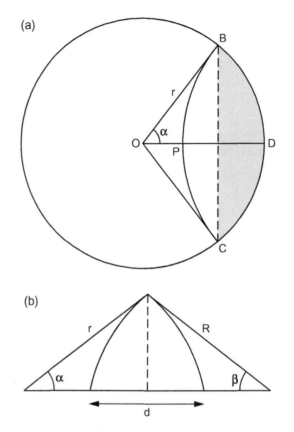

Figure 92.1(a,b). Geometry for the portion of the Sun covered by the Moon

  (iii) The relative speed between the Sun and Moon is assumed to be constant.

  (iv) The Sun's disk is assumed to be uniformly bright.

The basic parameters and diagrams for the model are kept the same as in the Möllman-Vollmer paper. At an arbitrary point of time in an arbitrary eclipse (e.g., partial, annular, or total), the area of the double segment *PBDCP* is found from elementary trigonometry applied to figure 92.1 to be

$$A = r^2(\alpha - \sin\alpha\cos\alpha) + R^2(\beta - \sin\beta\cos\beta). \tag{92.1}$$

The law of cosines relates R, r, α and d, where $d = PD$. Thus,

$$R^2 = r^2 + (R + r - d)^2 - 2r(R + r - d)\cos\alpha, \tag{92.2}$$

from which

$$\cos \alpha = \frac{r^2 + (R + r - d)^2 - R^2}{2r(R + r - d)}. \tag{92.3}$$

By symmetry, the corresponding expression for $\cos \beta$ may be found by interchanging r and R in the above formula, so that

$$\cos \beta = \frac{R^2 + (R + r - d)^2 - r^2}{2R(R + r - d)}. \tag{92.4}$$

Obviously, an important quantity in this context is the degree of obscuration p, which is defined as the ratio of the portion of the Sun covered by the Moon to the area of the Sun's disk, πr^2. This is, therefore,

$$p = \frac{A}{\pi r^2} = \frac{r^2(\alpha - \sin \alpha \cos \alpha) + R^2(\beta - \sin \beta \cos \beta)}{\pi r^2}. \tag{92.5}$$

In terms of the maximum illuminance L_{max}, the normalized illuminance is simply expressed as

$$\frac{L}{L_{max}} = 1 - p. \tag{92.6}$$

Since we are dealing with *relative* speed (v), it is reasonable to consider the Sun to be fixed during the eclipse and the Moon to be moving at speed v. The authors utilize a concept from particle physics known as the *impact parameter* (a); this is the minimum distance between the centers of the lunar and solar disks (as viewed, of course, in projection, by the observer). In terms of the distance between the centers *in the direction of lunar motion* (see figure 92.2), denoted by s, it is clear that

$$(R + r - d)^2 = a^2 + s^2, \tag{92.7}$$

which means that

$$d = R + r - (a^2 + s^2)^{1/2}, \tag{92.8}$$

(because $d < R + r$).

If the first contact between the disks occurs when $s = s_0$, say, the total distance traveled between the first and last contacts is, by symmetry, $2s_0$.

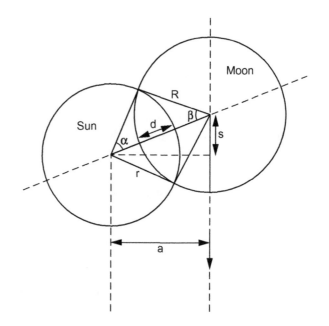

Figure 92.2. Basic geometry for solar obscuration during an eclipse

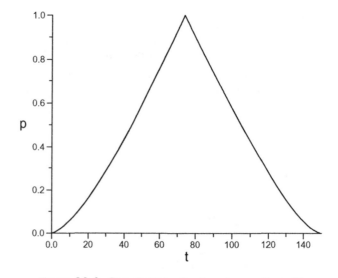

Figure 92.3. The degree of solar obscuration $p(t)$

Furthermore, if the time interval between these contacts is $2T$ (again allowing for the symmetry of the eclipse), then $v = s_0/T$. At maximum obscuration (whether total or partial), s can be represented as a function of time t by the simple relation

$$s(t) = s_0\left(1 - \frac{t}{T}\right)$$

(92.9)

since $s(T) = 0$. This now permits us to express all the parameters d, α, β, and, of course, the illuminance L as functions of t.

It is now possible to calculate the illuminance, in principle, for any type of eclipse; for an annular eclipse there are four points of contact between the two disks, and these can be incorporated into the calculations. In the spirit of this book, however, we concentrate on a simple analytical solution, namely one for which the eclipse is total, but for only a fleeting moment (the duration of totality is zero!), i.e., $R = r$ and $a = 0$. Now $\alpha = \beta$, and therefore it follows from equation (92.5) that

$$p = \frac{2(\alpha - \sin \alpha \cos \alpha)}{\pi},$$

(92.10)

and from (92.8)

$$d(t) = 2R - s(t),$$

(92.11)

where now $s_0 = 2R$ so that

$$d(t) = 2R\frac{t}{T}.$$

(92.12)

Hence, from equation (92.3) or (92.4),

$$\cos \alpha = 1 - \frac{t}{T}; \quad \sin \alpha = \left[2\frac{t}{T} - \left(\frac{t}{T}\right)^2\right]^{1/2},$$

(92.13)

so finally it follows that

$$\frac{L(t)}{L_{\max}} = 1 - p(t)$$

$$= 1 - \frac{2}{\pi}\left\{\arccos\left(1 - \frac{t}{T}\right) - \left(1 - \frac{t}{T}\right)\left[2\frac{t}{T} - \left(\frac{t}{T}\right)^2\right]^{1/2}\right\}.$$

(92.14)

The related function $p(t)$ is plotted in figure 92.3 for an arbitrary (but not unreasonable) value for the midpoint T of 75 minutes (which is now the time from first contact to totality). As seen, the graph of p in the interval $75 \leq t \leq 150$ is symmetrical about the line $t = 75$.

The authors of the original paper found very good agreement between their generalization of this analytical model and the data from several eclipses.

At the end . . .

Since the questions in this book have been prompted by things I have noticed and wondered about on my many walks, both long and short, over the years, it seems to me to be appropriate to conclude with some items that have to do specifically with walking, and endings . . . So without further ado, let's begin the final section with a question about perambulation!

Q.93: How can you model walking?

No one denies the obvious and numerous health benefits of walking. For a start, it is good for your heart. One recent study indicates that walking at a moderate pace (3 mph) for up to 3 hours a week—or 30 minutes a day—can cut the risk of heart disease in both men and women by a significant amount, perhaps as much as 40%. This is comparable to the benefit derived from aerobics, jogging, or other vigorous exercise. Related to this is the general knowledge that walking can help improve circulation, breathing, and the immune system, can control weight, and can help to prevent and control diabetes, among other things. So we're all agreed, right, that walking is good exercise. In question 33, I alluded to my average speed on my early morning walks. It's about 4.5 mph, so that puts me squarely in the category of *speed walking*. What is speed walking? Simply put, speed walking—sometimes called power walking, fitness walking, health walking, exercise walking, or striding—is walking very fast without running. Arms are swung in pace with the stride, and one foot is on the ground at all times. Your stride is slightly longer and considerably quicker than in a leisurely stroll. Speed walkers generally walk at a pace of 3.5 to 5.5 miles per hour. This increases my caloric burn rate without the joint-jarring effects of jogging. But *race-walkers* travel even faster, from 5 to 9 mph, although some competitive race-walkers can cover a mile in as little as six minutes. The object of race-walking, according to proponents, is to move your body ahead as quickly as possible without running and to avoid

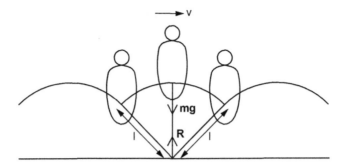

Figure 93.1. A simplistic model of walking (redrawn from Alexander (1996))

the up–down motions of regular walking. With all this wonderful information in mind. let's ask a question before developing models of walking. The
question is a biomechanical one, namely, *At what speed does walking turn into
running*, i.e., *How fast can we walk?*

A PRELIMINARY MODEL

We can use the following information based on a simplistic model (see figure
93.1): assume that the center of mass of the body moves in a series of circular
arcs, the radius of which is the length of the pivoting leg l, and whose center is
on the ground at the foot, while the body moves forward in a straight line
with speed v. We can use Newton's second law to balance the weight of the
body of mass m, the reaction R of the road on the walker, and the corresponding centripetal force mv^2/l.

This latter force is supplied by the weight $W = mg$ of the walker, and by
Newton's third law (action and reaction are equal and opposite) this is equal
and opposite to $R > 0$; when $R = 0$, running occurs. Thus, the difference

$$mg - \frac{mv^2}{l} = R \geq 0, \qquad (93.1)$$

where $R > 0$ for walking, and $R = 0$ for running (because the pivot leg is temporarily off the ground; if this is more than temporary then flight has been
achieved!) Hence, for walking we obtain the rather crude inequality

$$v < v_{max} = \sqrt{gl}. \qquad (93.2)$$

(In fact, the dimensionless number v^2/gl [<1 for walking] is called the
Froude number, after a Victorian engineer who analyzed wave resistance to

ships.) With $g \approx 10\,\text{m/s}^2$ and l in the range 0.8 to 0.9 m, we find that v_{max} lies in the range 2.8 to 3.0 m/s (9.2 to 9.8 fps, or 6.3 to 6.7 mph), about right for race walking. Notice that on the basis of this model we can " justify" the use of the simple pendulum as a first approach to "gait analysis": since in this model the pivot leg is assumed to move in a circular arc, the angular speed $\omega = v/l$, and we may define a "period" of swing as $T = 2\pi/\omega$. But

$$\omega = \frac{v}{l} < \frac{v_{max}}{l} = \frac{\sqrt{gl}}{l} = \sqrt{\frac{g}{l}} \equiv \omega_{max}. \tag{93.3}$$

Therefore,

$$T > T_{min} = \frac{2\pi}{\omega_{max}} = 2\pi\sqrt{\frac{l}{g}}, \tag{93.4}$$

a formula well known as the period (T_{min} here) of a simple pendulum.

AN IMPROVED MODEL

If you watch people walk, you will notice that, in general, there is a small amplitude oscillation of the top of the walker's head (in particular) associated with the regular horizontal motion. Research on the human gait suggests that the path $y(x)$ of the center of mass of the walker can be modeled by the curve $y = k + a \sin wx$ as shown in figure 93.2, where the quantity w is related to the stride length s by the equation $w = 2\pi/s$ because adjacent maxima must be separated by angular measure 2π (note that w and ω refer to separate quantities).

Suppose that ρ is the radius of curvature of the arc at the maxima; clearly at such points

$$y'(x) = aw \cos wx = 0, \tag{93.5}$$

and

$$y''(x) = -aw^2 \sin wx = -aw^2. \tag{93.6}$$

Therefore, at a maximum, the curvature

$$\kappa = \frac{|y''|}{(1 + y'^2)^{3/2}} = aw^2, \tag{93.7}$$

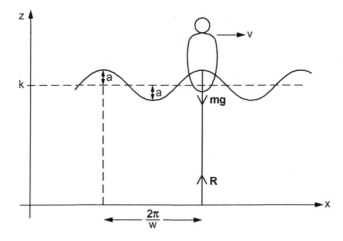

Figure 93.2. The improved model (redrawn from Alexander (1996))

and so

$$\kappa^{-1} = \rho = (aw^2)^{-1}. \tag{93.8}$$

Invoking Newton's second law of motion (see (93.1)) we get

$$mg - R = \frac{mv^2}{\rho} = mav^2w^2, \tag{93.9}$$

so that for walking, the following inequality holds:

$$R = mg - mav^2w^2 > 0, \tag{93.10}$$

implying that

$$v < v_{\max} = \sqrt{\frac{g}{a}} \frac{1}{w}; \tag{93.11}$$

this is our second inequality involving walking speed v. If we retain both in-equalities, they imply results consistent with what is observed in the gait of speed walkers, namely that for the leg in contact with the ground there is a marked *backward* rotation of the hip, while the other side of the walker's body "drops down." The former implies a small effective *extension* of the leg, so increasing l in inequality (93.2) implies that $v_{\max} = \sqrt{gl}$ also increases. The

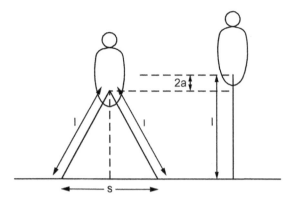

Figure 93.3. Incorporation of stride length (redrawn from Alexander (1996))

other statement implies that the amplitude a decreases, also resulting in an increase in v_{max} via the second inequality.

Noting that the maxima occur when one leg only is on the ground, and making the reasonable assumption that minima occur when the body weight is carried equally by both feet, then according to the sketch in figure 93.3,

$$2a = l - \left(l^2 - \frac{s^2}{4} \right)^{1/2}.$$

(93.12)

Using this result in the inequality (93.11), we can readily show that

$$v^2 < v_{max}^2 = \frac{g}{\pi^2} \frac{s^2}{2l - (4l^2 - s^2)^{1/2}} = \frac{g}{\pi^2} [2l + (4l^2 - s^2)^{1/2}].$$

(93.13)

Obviously $s < 2l$, and it is clear that as the stride length s decreases, v_{max} increases, or equivalently, if the maximum speed is to increase, the stride length must decrease, something that by this stage is perhaps intuitively obvious. In practice, since s is smaller than l, the value of v_{max} according to (93.13) does not vary significantly; if $l = 0.8$ m and $s = 0.6$ m, then $v_{max} \approx 1.8$ m/s (about 4 mph; within the range for speed walking), and if, for the same value of l, $s = 0.3$ m, v_{max} is almost the same, both significantly lower than in the cruder model.

Q.94:   How "long" is that tree?

Although this question might seem more appropriate for the section entitled "In the forest," I wanted to include it here because it concerns an ending of

Question 94. A "fractal" oak tree, in a village near Reading, England.
(See the description in question 94 for more details.)

sorts. It involves a favorite tree of mine, one that I call a "fractal oak," located very near to where my mother used to live in England. It was situated just a few minutes walk away, round the corner from her little apartment. I used to photograph it whenever I visited her, so I have pictures of it at various seasons of the year, though my favorite one was taken in winter, when the tree was "naked," and its impressive branching structure was clearly visible. My mother died early in March 2007, so in that sense it is an ending, for I doubt that I will ever have reason to visit this tree again. The color version of this picture was published in November 2004 on the Earth Science Picture of the Day site (EPOD; *http://epod.usra.edu/*), with the following caption:

> This stately old oak was photographed in the village of Woodcote, England (near Reading), in March of 2003. It's about 50 ft high (15 m), and the smallest twigs I could discern were about 1/2 inch (1.25 cm) long. I call this image "Fractal Oak" because of its seemingly statistical self-similarity over about three orders of magnitude. A fractal is essentially a fragmented geometric shape that can be subdivided in parts, each of which is, at least approximately, a small scale copy of the whole. Cloud edges and coastlines are just two examples of the many phenomena that fit this definition.

From examining this picture, I'll suppose for simplicity that each branch bifurcates into two smaller branches with equal radii: if $r_1 = r_2$ the statement

$r_0^3 = r_1^3 + r_2^3$ is equivalent to $r_1 = 2^{-1/3} r_0 \approx 0.794 r_0$. In other words, we'll assume that every branch in my now generic "oak tree" arises from an equal bifurcation of a larger branch. Such bifurcations are common in the vascular system of humans and other animals, and so it is interesting to make some comparisons between the two. For comparison with a typical animal vascular system, it transpires that a typical capillary radius is about 5 microns. It is therefore of interest to calculate how many bifurcations from a given primary vessel (or trunk, in the case of the tree) are required to reach this radius. Therefore, the number of bifurcations n from, say, an aorta of radius r_0 that will result in such a capillary must satisfy the equation

$$(0.794)^n r_0 = 5 \times 10^{-6} \text{ m.} \tag{94.1}$$

For an animal such as a dog, $r_0 \approx 0.5 \text{ cm} = 5 \times 10^{-3} \text{ m}$, so

$$(0.794)^n = 10^{-3}, \text{ i.e., } n = \frac{-3}{\log_{10}(0.794)} \approx 30. \tag{94.2}$$

It follows from this that the total number of vessels in the system after 30 bifurcations is $2^{30} \approx 10^9$. Let's just remind ourselves of some properties of geometric series. The following series of $n+1$ terms

$$a + ar + ar^2 + \cdots + ar^n$$

sums to

$$S_n = a \frac{r^{n+1} - 1}{r - 1}, \quad r \neq 1. \tag{94.3}$$

(More accurately, we should write S_{n+1} for this sum, but S_n reminds us that there are n bifurcations.) The ratio of the last term in the series to its sum is just $r^n(r-1)/(r^{n+1}-1)$; for the present discussion $r=2$, and this ratio is approximately $1/2$ for large enough values of n. Clearly, for $n=30$ this is extremely accurate. Therefore, the last bifurcation results in half this number of vessels, i.e., 5×10^8, which is not far from the estimate of 1.2×10^9 given in the literature (see the reference to Rosen). And as pointed out by Rosen, the agreement becomes even closer when one realizes that observationally, it is probably difficult to distinguish vessels in the final bifurcation from those arising from several preceding ones (i.e., the 29th or even 28th bifurcation).

Now let's try and obtain an estimate of the total *length* of the branching system. In the absence of much physiological information, and in the spirit of a "back of the envelope" calculation, we suppose that the trunk (or aorta) has length L_0, and that each bifurcation k produces twice as many branches, each of length $L_k = \alpha L_{k-1}$, where it is assumed that $0 < \alpha < 1$. Then the total length of the system after n bifurcations is

$$
\begin{aligned}
L_n &= L_0 + 2\alpha L_0 + (2)^2\alpha L_1 + (2)^3\alpha L_2 + \cdots + 2^n\alpha L_{n-1} \\
&= L_0[1 + 2\alpha + (2\alpha)^2 + (2\alpha)^3 + \cdots + (2\alpha)^n] \\
&= L_0 \frac{(2\alpha)^{n+1} - 1}{2\alpha - 1}, \quad \alpha \neq \frac{1}{2}.
\end{aligned}
\tag{94.4}
$$

I estimated that the narrowest twig corresponded to a *radius* of about 1 mm, and with an estimate of $r_0 \approx 1$ m, we have from (94.1)

$$
(0.794)^n r_0 = 10^{-3} \text{m},
$$

i.e.,

$$
n = \frac{-3}{\log_{10}(0.794)} \approx 30,
\tag{94.5}
$$

again, interestingly, the same number of bifurcations as for the dog, if my observations of the tree are reasonably accurate. Turning to the calculation of length we find that if I choose $L_0 = 3$ m for the trunk and $\alpha = 2/3$,

$$
L_{30} = 3\frac{(4/3)^{31} - 1}{1/3} \text{ m} \approx 7 \times 10^4 \text{ m} = 70 \text{ km},
\tag{94.6}
$$

whereas if $\alpha = 7/8$, then

$$
L_{30} = 3\frac{(7/4)^{31} - 1}{3/4} \text{ m} \approx 10^8 \text{ m} = 10^5 \text{ km}.
\tag{94.7}
$$

That would be a lot of tree!

Q.95:   What are those "rays" I sometimes see at or after sunset?

These rays, often alternating bands of light and dark, are known as *crepuscular rays*, or just sunbeams. The word "crepuscular" is derived from the Latin word *crepusculum*, meaning twilight. The dark bands are shadows of clouds, and

Question 95. Crepuscular rays. The dark bands are shadows of clouds (or sometimes mountains) cast on the sky. The bright bands are sunlight, contrasting with these shadows, and made visible by atmospheric particles scattering light into the observer's eyes.

the bright bands are sunlight, contrasting with these shadows. Crepuscular rays are most frequently seen emanating from the Sun's location, even after it has set, but on the particular evening described below, only their extension in the *east* was visible; these are known as *anticrepuscular* rays. They appear to converge at the antisolar point (the point diametrically opposite the Sun); but since they are parallel this is merely an effect of perspective. In this respect, they have much in common with the mountain shadows discussed in question 85; indeed, mountain shadows can be thought of as an example of anticrepuscular rays.

If a ray shines through a gap in the clouds, it is bright, and if a cloud casts a shadow on the sky, the "ray" appears to be dark by contrast. In either case, as pointed out in an article by Dave Lynch (see the references), scattering particles—such as dust, rain, snow, or even just air molecules—are required to make the ray visible to an observer. He identifies some common features of these rays, and finds that many aspects of their visibility are independent of the scattering mechanism, and can be explained by the geometry of shadows, in particular, in terms of the line-of-sight distance D through the umbral shadow (see question 85). The observant reader may have noticed some of these features, which include:

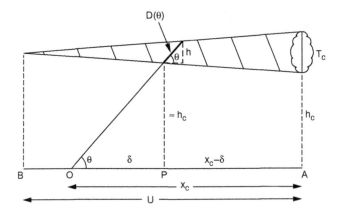

Figure 95.1. Basic geometry of cloud location and its umbral shadow

(i) Rays are most common when the Sun is low *and* when the air is clear.

(ii) The greatest ray visibility (or contrast) occurs within 15° of the Sun and the antisolar point, and, in the latter case, reach their maximum visibility within a few degrees of this point.

(iii) Anticrepuscular rays are usually associated with large clouds, either overhead or near the antisolar point; they are seldom associated with small clouds unless they are near the antisolar point.

(iv) Maximum *intensity* contrast occurs near the Sun, and maximum *color* contrast occurs near the antisolar point.

We can use formula (36.1) in question 36 to gauge the approximate length U of the umbral shadow of a cloud. In what follows, T_c is the vertical thickness of the cloud, the base of which is at height h_c; x_c is the horizontal distance of the cloud from the observer. Rewriting that formula in terms of the notation of figure 95.1, we find that

$$U \approx \frac{T_c s}{2R}, \tag{95.1}$$

where $s \approx 9.3 \times 10^7$ miles is the approximate Earth–Sun distance, and $R \approx 4.3 \times 10^5$ miles is the radius of the Sun. Therefore, $U \gtrsim 100 T_c$ miles, so for a large cloud of height 1 mile, the shadow length is more than 100 miles. Because of this $U{:}T_c$ ratio of about one hundred, the taper of the umbral shadow is small, so we see from figure 95.1 that $h_c \approx \delta \tan \theta$, where δ is the distance OP. For the same reason, the vertical thickness of the umbra at that point is $h \approx D(\theta) \sin \theta$, where the distance $D(\theta)$ is a measure of the contrast

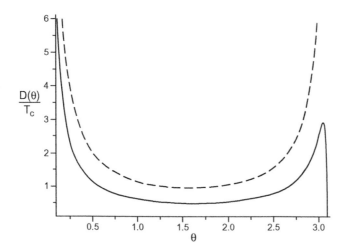

Figure 95.2. Graphs of $D(\theta)/T_c$ for two sets of data (x_c, h_c, u)

of the ray as a function of the observer's angle of sight θ, measured from the ground. By similar triangles,

$$\frac{h}{T_c} \approx \frac{U - (x_c - \delta)}{U} = 1 - \frac{x_c - \delta}{U}.$$

Therefore,

$$D(\theta) \approx h \csc \theta \approx T_c \left(1 - \frac{x_c - \delta}{U}\right) \csc \theta = T_c \left(1 - \frac{x_c - h_c \cot \theta}{U}\right) \csc \theta.$$

$$(95.2)$$

This formula is derived for the Sun positioned close to the horizon, i.e., near sunset (or sunrise). Figure 95.2 illustrates the function $D(\theta)$ according to equation (95.2).

In units of miles, the upper (dashed) curve is for the arbitrarily chosen values $x_c = 10$, $h_c = 2$, and $U = 200$. For the lower (solid) curve, $x_c = 70$, $h_c = 3$, and $U = 140$. The angle θ is expressed in radians. Near $\theta = \pi$, the graphs turn down steeply because of the singular behavior of $D(\theta)$ as θ approaches π from below; $\csc \theta \to \infty$ and $\cot \theta \to -\infty$ in this limit. Of course, there is corresponding (positive) singular behavior as $\theta \to 0$ through positive values. Both these cases are consistent, to this level of approximation, with features (ii)–(iv) noted above.

But why is *this* the penultimate question, dealing as it does with crepuscular rays? The answer is "because of twilight" . . . I placed this question here

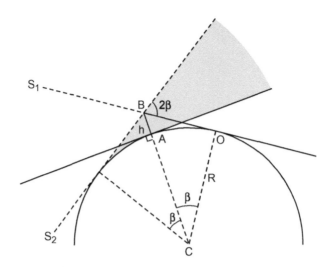

Figure 96.1. Geometry for the twilight calculation

because you, the reader, are in the twilight of the book! But to illustrate that no metaphor is ever perfect, I should remind you that these rays are frequently visible during the day, though, as we have seen, the contrast between light and shadow is usually best appreciated around sunrise or sunset. I will finish this question with a personal example. On vacation in Yellowstone National Park in the middle of June, my wife and I had joined a gathering of about twenty people on the side of the road who were studying distant wildlife (wolves and bears) through powerful telescopes and binoculars. It was just after sunset, and all of a sudden, I gazed up into the eastern sky and saw some magnificent anticrepuscular rays. I drew attention to them, and most of the group was quite impressed with this wonderful display before returning to the task at hand!

Q. 96: How can twilight help determine the height of the atmosphere?

According to Götz Hoeppe, in his book *Why the Sky Is Blue*, the eleventh-century mathematician Abu'Abd Allah Muhammad ibn Muadh suggested that the "angle of dip" of the Sun below the horizon at twilight can be used to determine the height of the atmosphere. Of course, we know that the atmosphere does not suddenly end, and as noted below, in modern times, there are several different ways of defining this height. Based on the duration of twilight, Abu'Abd Allah's estimate of this angle, 2β in figure 96.1, was 18°.

Evening twilight commences, for an observer at O, with the Sun setting below the horizon (position S_1 in the figure), and as it sets further, progressively higher atmospheric layers enter the Earth's shadow. Eventually, when the Sun is at position S_2, the "top" of the atmosphere (point B) is in shadow, and this marks the end of astronomical twilight and the beginning of night. From the figure it is clear that if the radius of the Earth, CO, is R, then

$$\cos \beta = \frac{R}{R+h},\qquad (96.1)$$

from which it follows that

$$h = R(\sec \beta - 1).\qquad (96.2)$$

Using the value $R \approx 6370$ km, and $\beta = 9°$ (or $\pi/20$ radians), equation (96.2) yields a value for h of approximately 79 km (49 miles). And this brings us back to two definitions of atmospheric height (there are others, but these two are the most relevant for this question). NASA's definition is that "outer space" starts at a height of 50 miles, whereas the standard international definition is that it starts at a height of 100 km (62 miles). All told, Abu'Abd Allah's estimate of β, combined with an accurate value of R provides us with a very reasonable value of h!

Spheroid, prolate ellipsoid

A *spheroid* is a surface in three dimensions obtained by rotating an ellipse about one of its principal axes. If the ellipse is rotated about its major axis, the surface is a prolate ellipsoid (similar to the shape of a rugby ball or American football).

Hyperbolic tangent function (tanh *x*)

This function is defined as follows (see figure G.1):

$$\tanh x = \frac{e^x - e^{-x}}{e^x + e^{-x}} = \frac{e^{2x} - 1}{e^{2x} + 1}, \tag{A1.1}$$

$$\lim_{x \to \infty} \tanh x = 1 \text{ and } \lim_{x \to -\infty} \tanh x = -1. \tag{A1.2}$$

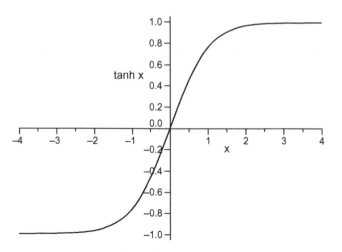

Figure G.1. tanh x

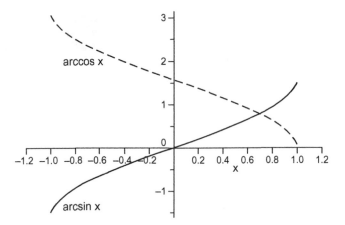

Figure G.2. arcsin x (solid line) and arccos x (dashed line)

Inverse sine and inverse cosine functions

These are defined as follows (see figure G.2):

$$\text{For } y \in \left[-\frac{\pi}{2}, \frac{\pi}{2}\right], y = \arcsin x \text{ if } x = \sin y, \quad \text{and} \qquad (A1.3)$$

$$\text{For } y \in [0, \pi], y = \arccos x \text{ if } x = \cos y. \qquad (A1.4)$$

Alternatively,

$$\arcsin x = \int_0^x \frac{dt}{\sqrt{1 - t^2}}, \quad |x| < 1. \qquad (A1.5)$$

Since

$$\arcsin x + \arccos x = \frac{\pi}{2}, \qquad (A1.6)$$

a corresponding integral form for arccos x follows directly.

Bessel function of the first kind of order one

The Bessel function of the first kind of order one ($J_1(x)$, the only Bessel function referred to in this book), like all Bessel functions, can be expressed as an infinite series, but all that we need to be aware of for question 91 is that for

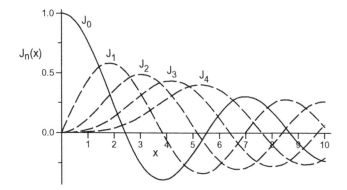

Figure G.3. Graphs of Bessel functions of the first kind, J_n (x), n = 0, 1, 2, 3, 4

large values of x, $J_1(x)$ behaves somewhat like a damped cosine function. Specifically, as $x \to \infty$,

$$J_1(x) \sim \sqrt{\frac{2}{\pi x}} \cos\left(x - \frac{3\pi}{4}\right), \tag{A1.7}$$

where for our purposes the "$\sim$" can be interpreted as the phrase "behaves like." The reason that the graph in question 91 doesn't look much like the J_1 function in figure G.3 is that it is a graph of, essentially, $[2J_1(x)/x]^2$, and so is especially large for smaller values of x, and the square exponent exacerbates that effect. Furthermore, that graph is symmetric about $x = 0$ because x is a measure of the angle from the center of the coronal pattern. Nevertheless, the oscillatory behavior associated with the J_1 function is readily apparent. Bessel functions of orders 0, 2, 3, and 4 are included for comparison in figure G.3.

Q.1:   You are looking at a single bright rainbow (the primary bow). Which color is on the top side of the arch?

Answer: Red. The violet light is deviated more than the red light as a result of being refracted and reflected inside the drop. Consequently, as viewed by the observer, the red portion of the rainbow appears closer to the Sun (in an angular sense).

Q.2:   How many colors do you think there are in a typical rainbow?

Answer: Well, it isn't seven! The old ROYGBIV or VIBGYOR are useful mnemonics for the rainbow colors, but, in fact, there are, mathematically at least, an infinite number of "colors" since the visible spectrum is a continuum. Of course, for perceptual reasons, the color resolution of the human eye/brain connection is not infinite, so we do tend to interpret what we see as a discrete set of colors. But has anyone *ever* seen *indigo*?

Q.3:   Suppose you see a double rainbow; which color is on the top side of the upper arch (the secondary bow)?

Answer: Violet. The reason for this is that the additional reflection inside the drop reverses the order of the colors compared with the primary bow.

Q.4:   Is the region between the two bows typically darker, brighter, or the same as the surrounding sky?

Answer: Usually it is darker, because raindrops between the bows do not refract and reflect light to the observer's eye. This angular region is called *Alexander's dark band,* and is particularly noticeable by contrast if the bows are especially bright.

Q.5:   How about the region below the primary bow?

Answer: Typically it is brighter, because light from raindrops causing the rainbows is scattered into the observer's eye. The rainbows themselves are a result of a concentration of light in a small angular region; outside of this small region, light is scattered into directions both above and below the bows.

Q.6:   Have you ever seen anything else closely associated with the primary bow?

Answer: Yes, I have; good question! Sometimes two or more pastel colored fringes can be seen below the upper regions of the primary. They were a source of considerable scientific interest by the mid-eighteenth century, because neither Newton's nor Descartes' theories of the rainbow could account for them. They are, in fact, a consequence of the wavelike nature of light, namely interference, and are an integral part of the rainbow. Indeed, Lee and Fraser, in a delightful phrase found in *The Rainbow Bridge*, note that "the supernumerary rainbows proved to be the midwife that delivered the wave theory of light to its place of dominance in the nineteenth century."

Q.7:   On a sunny day, are the edges of your shadow sharp, diffuse, or some combination of both?

Answer: The parts of my shadow closest to me are sharp, but my head can appear somewhat fuzzy(!). This effect is particularly noticeable for tree shadows, especially during the autumn or winter when the leaves have dropped and when the sun is not too high. The "fuzzy" aspect is due to the penumbral (or partial) shadow; a consequence of the sun being an extended source of light (see questions 36 and 85 for more details about shadows).

Q.8:   Over what typical timescale does a small cumulus cloud maintain its original shape (i.e., from when you first looked at it)?

Answer: This is a difficult question to answer with any precision, because atmospheric conditions can vary so much, but I generally suggest that the shape will have changed noticeably within a five-minute period, based on my observations.

Q.9:   Estimate the size (diameter) of water droplets in (i) a heavy downpour, and (ii) fog.

Answer: (i) 1–2 mm; (ii) 0.01–0.1 mm. These can be crudely guessed at from the watery deposits on one's spectacles (assuming one wears them) in the rain or in the fog!

Q.10:   To the nearest order of magnitude, how many light waves would fit across your fingernail?

Answer: A fingernail is about 1 cm in size, and the visible spectrum has waves of wavelength (λ) in the range 400 nm (violet) to 700 nm (red). For simplicity, therefore, taking a typical value of $\lambda \approx 500$ nm, or 5×10^{-5} cm, we have that of order 10^4 such waves would cover my fingernail.

Q.11:   How long is a typical sound wave associated with human speech?

Answer: I included this question for direct comparison with question 10. The normal frequency (v) range of human speech, much narrower than that for hearing, is from about 500 to 2500 Hz. If c is the speed of sound at room temperature ($c \approx 330$ m/s), then since $\lambda = c/v$, the wavelength ranges from as short as about 10 cm to about 1 m, rounding to the nearest order of magnitude. The corresponding wavelength range for human hearing is about 1 cm to 10 m.

Q.12:   What is the (eastward) rotation speed of the Earth at the equator? At the poles?

Answer: The Earth rotates once, by definition, in a day, so its speed v is one circumference per day, i.e., $v = 2\pi R$ per day (R being the radius of the Earth), or $v \approx (6 \times 4000)/24 = 1000$ mph, or about 1600 km/hr at the equator. At the poles it is zero!

Q.13:   At what latitude is the rotation speed the (arithmetic) mean of the correct answers to the two parts of the previous question?

Answer: At a latitude θ of 60°N (or S) since then the effective "radius" of the Earth is $R \cos \theta = R/2$. (Or think about the trigonometry of the unit circle.)

Q.14:   How far away is the horizon if you are standing at the beach looking out to sea?

Answer: This is a test—you should know the answer from question 39! Neglecting the effects of atmospheric refraction, as the simple geometric argument of question 39 shows, if the observer's eyes are h ft above the ground, the distance d in miles to the horizon is $d \approx \sqrt{1.5h}$ to a good approximation. And we noted there that if $h = 6$ ft, then $d \approx 3$ miles; if $h = 150$ ft, then $d \approx 15$ miles. If atmospheric refraction is included, the distance seen by an observer near the ground is extended by about 9%.

Q.15: In the middle of the day, you are looking at two similar hills, except that one is more distant than the other. The one that appears a bit darker is (i) the nearer one; (ii) the more distant one; (iii) both appear equally dark, or (iv) take off your sunglasses, silly!

Answer: (i) The nearer one, because the effects of Rayleigh scattering (which makes the daylight sky appear so blue) are less than for more distant hills. This is because the degree of scattering is obviously smaller for the nearer hills, there being less air in the line-of-sight of the observer.

In question 83, a simple equation describing the growth of a "tree tumor" was derived. This equation appears in many different contexts, most of which are outside the focus of this book. However, one such context seems appropriate to mention: the *cooling* of the many cups of coffee I drank while writing this book. We consider the statement of *Newton's law of cooling*: the temperature $T(t)$ of an object will change at a rate proportional to the difference between its temperature and that of its surroundings, T_s. It is entirely reasonable to suppose that the rate of change $T'(t)$ is a continous function (i.e., $T(t)$ is a differentiable function of time t), and that the gain or loss of heat from the object to the enclosure is negligible, so that T_s is constant. This would be the case, for example, with a hot cup of coffee on a table in the kitchen (and also for a turkey in an oven maintained at a constant temperature). We define $T(t) = T_0$.

Expressed in mathematical terms, we have that

$$\frac{dT}{dt} = -K(T - T_s), \tag{A3.1}$$

K being a positive constant of proportionality. Why positive? Well, let's think about it: if the object is *hotter* than its surroundings ($T > T_s$), the right-hand side of this little equation is $-K(T - T_s) < 0$, so according to the equation the object cools as expected. If the object is *cooler* than its surroundings ($T < T_s$), then the right-hand side is positive, indicating its temperature *increases* with time. So both possibilities are covered by the same equation; now *that* is cool! Since the equation is a separable one, we may readily integrate it:

$$\int \frac{dT}{T - T_s} = -K \int dt,$$

i.e., $\ln|T - T_s| = -Kt + \ln D, \tag{A3.2}$

where by setting $t=0$ in the above equation we see that $D=|T_0-T_s|>0$. Therefore,

$$|T - T_s| = |T_0 - T_s|e^{-Kt}. \tag{A3.3}$$

Consider first the object to be cooling, so $T_0 \geq T > T_s$ for all finite times and $\lim_{t \to \infty} T = T_s$. Then (A3.3) becomes

$$T = T_s + (T_0 - T_s)e^{-Kt}, \tag{A3.4}$$

corresponding to exponential decay toward T_s. If the object is heating up, then $T_s > T \geq T_0$ and $\lim_{t \to \infty} T = T_s$ as before, and it is easily seen that equation (A3.4) still applies.

A *short cut*: If we make the change of variable $\theta = T - T_s$ in equation (A3.1), the resulting equation is

$$\frac{d\theta}{dt} = -K\theta, \tag{A3.5}$$

with solution

$$\theta = \theta_0 e^{-Kt}, \tag{A3.6}$$

and in terms of the original variable T this is exactly the form of equation (A3.4). Why didn't I mention that at the beginning? Because we wouldn't have had so much fun thinking about the meaning of the solution, would we? Now go back and look at question 83 again!

What follows below is obviously only a partial list of patterns that an attentive observer might see on a "nature walk." Some of these patterns have, of course, been noted in the body of this book. For example, basic two-dimensional geometric shapes that occur (approximately) in nature can be identified:

- Waves on the surfaces of ponds or puddles expand as circles.
- Ice crystal halos commonly visible around the sun are generally circular.
- Rainbows have the shape of circular arcs.
- Tree growth rings are almost circular.

But there are many other obviously noncircular and nonplanar patterns:

- Hexagons: snowflakes generally possess hexagonal symmetry.
- Pinecones, sunflowers, and daisies (among other flora) have spiral patterns associated with the well-known Fibonacci sequence.
- Ponds, puddles, and lakes give scenes *approximate* reflection symmetry (depending on the position of the observer).
- Cross sections of various fruits also exhibit interesting symmetries.
- Spider webs have polygonal, radial, and spiral-like features.
- Long, bendy grass has a parabolic shape.
- Starfish exhibit pentagonal symmetry.
- The raindrops that scatter "rainbow" light into the eye of an observer essentially lie on cones with vertex at the eye.
- Cloud patterns, mud cracks, and cracks on tree bark can have polygonal patterns.
- Clouds can also form wavelike "billow" structures with well-defined wavelengths, just as ripples have around rocks in a swiftly flowing stream.
- In three dimensions, snail shells and many seashells and curled-up leaves are helical in shape and tree trunks are approximately cylindrical.

In view of these patterns, even (and perhaps especially) at an elementary level, many pedagogic mathematical investigations can be developed to describe such patterns, for example, estimation, measurement, geometry, functions, algebra, trigonometry, and calculus of a single variable. At the level of elementary geometry, basic examples might include

- The use of similar triangles and simple proportion
- A table of tangents to estimate the height of trees.
- Measuring inaccessible horizontal distances using congruent triangles

Simple proportion can again be used in estimation problems, such as

- Finding the number of blades of grass in a certain area, or the number of leaves on a tree.

And more geometric ideas appear when studying topics such as

- The relationship between the branching of some plants, such as sneeze-wort (*Achillea ptarmica*), and the Fibonacci sequence can be investigated.
- The related "golden angle" can be studied, and its occurrence on many plants (such as laurel) investigated.
- The angles subtended by the fist and the outstretched hand at arm's length can be estimated, and used to identify the location of sundogs (parhelia) and ice crystal halos on days with cirrus clouds near the sun.

Consequences of the problem of scale and geometric similarity can also be investigated. This applies, in particular, to the size of land animals; the relationship of surface area to volume, and its implications for the relative strength of animals. By considering (and constructing) cubes of various sizes, much insight can be gained about basic biomechanics in the animal kingdom, and much fun (and learning!) may be had by thinking about such questions as *why King Kong could not really exist*, and *why elephants are not just large mice*! Furthermore, simple ideas such as scale enable us to compare, at an elementary level, metabolism and other biological features (such as strength) in connection with pygmy shrews, hummingbirds, insects and African elephants to name but a few groups!

References

Acheson, D. J. (1990). *Elementary Fluid Dynamics*. Clarendon Press, Oxford, UK.

Adam, J. A. (2002). "Like a Bridge over Colored Water: A Mathematical Review of *The Rainbow Bridge: Rainbows in Art, Myth and Science*" by R. Lee, Jr., and A. Fraser, *Notices of the AMS*, 49, 1360–1371.

Adam, J. A. (2003). "Mathematical Models of Tumor Growth: From Empirical Description to Biological Mechanism." *Advances in Experimental Medicine and Biology: Mathematical Modeling in Nutrition and the Health Sciences* 537, 287–300 (Kluwer Academic/Plenum Publishers).

Adam, J. A. (Paperback edition 2006). *Mathematics in Nature: Modeling Patterns in the Natural World*. Princeton University Press, Princeton, NJ.

Adam, J. A. (2008). "Geometric optics and rainbows: generalization of a result by Huygens." *Applied Optics* 47: H11.

Adler, F. (1998). *Modeling the Dynamics of Life*. Brooks/Cole, Pacific Grove, CA.

Alexander, R. M. (1984). "Walking and Running." *American Scientist* 72: 348–354.

Alexander, R. M. (1996). "Walking and Running." *The Mathematical Gazette* 80, 262–266.

Ahrens, C. D. (2000). *Meteorology Today: An Introduction to Weather, Climate, and the Environment*. (6th edition). Brookes Cole, Pacific Grove, CA.

Austin, J. D. and Dunning, F. B. (1988). "Mathematics of the Rainbow." *The Mathematics Teacher* September: 484–488.

Baker, D. E. (2002). "A Geometric Method for Determining Shape of Bird Eggs." *The Auk* 119(4), 1179–1186.

Baldock, G. R. and Bridgeman, T. (1981). *The Mathematical Theory of Wave Motion*. Ellis Horwood, Chichester, UK.

Ball, P. (1999). *The Self-made Tapestry*. Oxford University Press, Oxford, UK.

Barber, N. F. (1969). *Water Waves*. Wykeham Press, London.

Bascom, W. (1980). *Waves and Beaches*. Anchor Press, New York.

Bohren, C. (1987). *Clouds in a Glass of Beer*. Wiley, New York.

Bohren, C. (1991). *What Light Through Yonder Window Breaks?* Wiley, New York.

Bohren, C. and Fraser, A. (1986). "At What Altitude Does the Horizon Cease to Be Visible?" *American Journal of Physics* 54:222.

Bohren, C. F. and Huffman, D. R. (1983). *Absorption and Scattering of Light by Small Particles.* Wiley: New York.

Bonner, J. T. (2006). *Why Size Matters.* Princeton University Press, Princeton, NJ.

Bowers, R. L., and Deeming, T. (1984). *Astrophysics, I: Stars.* Jones & Bartlett, Boston.

Boyer, C. B. (1987). *The Rainbow, from Myth to Mathematics.* Princeton University Press, Princeton, NJ.

Bryant, H. C. and Jarmie, N. (1974). Reprinted in *Light from the Sky.* (1980; Readings from *Scientific American*), pp. 66–74. W. H. Freeman, San Francisco.

Byrne, H. (April 1999). "The Role of Mathematics in Solid Tumour Growth." *Mathematics Today,* 48–53.

Callander, R. A. (1978). "River Meandering." *Annual Review of Fluid Mechanics* 10: 129–158.

Carter, T. C. (1968). " The Hen's Egg: A Mathematical Model with Three Parameters." *British Poultry Science* 9, 165–171.

Cowley, L. *http://www.atoptics.co.uk/*

Cowley, L., Laven P., and Vollmer, M. (2005). "Rings Around the Sun and Moon: Coronae and Diffraction." *Physics Education* 40(1), 51–59.

Curle, S. N. and Davies, H. J. (1968). *Modern Fluid Dynamics* (Volume 1). Van Nostrand Reinhold, Wokingham, UK.

Elmore, W. C. and Heald, M. A. (1969). *Physics of Waves.* Dover, New York.

Ehrlich, R. (1993). *The Cosmological Milkshake.* Rutgers University Press, Piscataway, NJ.

Ferguson, R. I. (1973). "Regular Meander Path Models." *Water Resources Research* 9:1079–1086.

Ferguson, R. I. (1976). "Disturbed Periodic Model for River Meanders." *Earth Surface Processes* 1: 337–347.

Fraser, A. B. (1972). "Inhomogeneities in the Color and Intensity of the Rainbow." *Journal of the Atmospheric Sciences* 29: 211–212.

Fraser, A. B. and Mach, W. H. (1976). "Mirages." Reprinted in *Light from the Sky* (1980; Readings from *Scientific American*), pp. 29–37., W. H. Freeman, San Francisco.

French, A. P. (1982). "How Far Away Is the Horizon?" *American Journal of Physics* 50:795.

Gallant, R. A. (1987). *Rainbows, Mirages and Sundogs.* MacMillan, New York.

Garland, T. H. (1987). *Fascinating Fibonaccis.* Dale Seymour Publications: Palo Alto, CA.

Greenler, R. G. (1980). *Rainbows, Halos and Glories.* Cambridge University Press, Cambridge, UK.

Greenler, R. G. and Mallmann, A. J. (1972). "Circumscribed Halos." *Science* 176: 128–131.

Herd, T. (2007). *Kaleidoscope Sky.* Abrams, New York.

Hoeppe, G. (2007). *Why the Sky Is Blue: Discovering the Color of Life.* Princeton University Press, Princeton, NJ.

Hoyt, D. F. (1977). "The Effect of Shape on the Surface–Volume Relationships of Bird's Eggs." *The Condor* 78, 343–349.

Humphreys, W. J. (1964). *Physics of the Air. Dover,* New York.

Isenberg, C. (1992). *The Science of Soap Films and Soap Bubbles.* Dover, New York.

Jeans, J. H. (1961). *Astronomy and Cosmogony.* Dover, New York.

Laven, P. *http://www.philiplaven.com/index1.html*

Laven, P. (2005). "How Are Glories Formed?" *Applied Optics* 44: 5675–5683.

Lee, R. L. and Fraser, A. B. (2001). *The Rainbow Bridge: Rainbows in Art, Myth and Science.* Pennsylvania State University Press, University Park, PA.

Lemons, D. S. (1997). *Perfect Form: Variational Principles, Methods, and Applications in Elementary Physics.* Princeton University Press, Princeton, NJ.

Leopold, L. B. (1997). *Water, Rivers and Creeks.* University Science Books, Sausolito, CA.

Leopold, L. B. and Langbein, W. B. (1966). "River Meanders." *Scientific American* 214: 60–70.

Livingston, W. C. and Lynch, D. K. (1979). "Mountain Shadow Phenomena," *Applied Optics,* 18, 265–269.

Lynch, D. K. (1978). "Atmospheric Halos." Reprinted in *Light from the Sky* (1980; Readings from *Scientific American*), pp. 38–46., W. H. Freeman, San Francisco.

Lynch, D. K. (1980). "Mountain Shadow Phenomena, 2: The Spike Seen by an Off-Summit Observer." *Applied Optics* 19, 1585-1589.

Lynch, D. K. (1987). "Optics of Sunbeams," *Journal of the Optical Society of America* A4, 609–611.

Lynch, D. K. (2005). "Turbulent Ship Wakes: Further Evidence That the Earth Is Round." *Applied Optics* 44, 5759–5762.

Lynch, D. K. and Livingston, W. (2004). *Color and Light in Nature* (2nd edition). Cambridge University Press, Cambridge, UK.

Mao, S. *http://www.jb.man.ac.uk/~smao/starHtml/stellarEquation.pdf*

Mayo, N. (1997). "Ocean Waves—Their Energy and Power." *The Physics Teacher* 35: 352–356.

Minnaert, M. G. J., (1993). *Light and Colour in the Outdoors.* Springer, New York.

Mollmann, K-P., and Vollmer, M. (2006). "Measurements and Predictions of the Illuminance During a Solar Eclipse." *European Journal of Physics* 27, 1299–1314.

Naylor, J. C. (2002). *Out of the Blue.* Cambridge University Press, Cambridge, UK.

Nussenzveig, H. M. (1977). "The Theory of the Rainbow." Reprinted in *Light from the Sky* (1980; Readings from *Scientific American*), pp. 54–65., W. H. Freeman, San Francisco.

Paganelli, C. V., Olszowka, A., and Ar, A. (1974). "The Avian Egg: Surface Area Volume and Density." *The Condor* 76, 319–325.

Pedgely, D. E. (1986). "The Tertiary Rainbow." *Weather* 41:401.

Pesic, P. (2005). *Sky in a Bottle.* The MIT Press, Cambridge, MA.

Preston, F. W. (1953). "The Shapes of Birds' Eggs." *The Auk* 70, 160–182.

Preston, F. W. (1968). "The Shapes of Birds' Eggs: Mathematical Aspects." *The Auk* 85, 454–463.

Preston, F. W. (1974). "The Volume of an Egg." *The Auk* 91, 132–138.

Roper, T. E. (1990). "Mathematics and the Motion of the Human Body." *The Mathematical Gazette* 74, 19–26.

Rosen, R. (1967). *Optimality Principles in Biology.* Butterworth, London.

Rothenberg, R. J. (1991). *Probability and Statistics.* Harcourt Brace Jovanovich, New York.

Shott, A. R., and Preston, F. W. (1975). "The Surface Area of an Egg." *The Condor* 77, 103–104.

Spilhaus, A. F. (1947). "Raindrop Size, Shape and Falling Speed." *Journal of Meteorology* 5, 108–110.

Stevens, P. S. (1974). *Patterns in Nature.* Atlantic–Little, Brown, Boston.

Stewart, I. (1999). *Life's Other Secret: The New Mathematics of the Living World.* Wiley, New York.

Taffe, W. J. (1978). "Mathematics in a Pumpkin Patch." *The Mathematics Teacher,* October, 603–607.

Tape, W. (1985). "The Topology of Mirages." *Scientific American* 252, 120–129.

Tape, W. (1994). *Atmospheric Halos.* American Geophysical Union, Washington, DC.

Tatum, J. B. (1975). "Egg Volume." *The Auk* 92, 576–580.

Tatum, J. B. (1977). "Area–Volume Relationship for a Bird's Egg." *The Condor* 79, 129–131.

Tayler, R. J. (1970). *The Stars: Their Structure and Evolution.* Wykeham Press, London.

Thompson, D'Arcy W. (1992). *On Growth and Form.* Dover, New York.

Todd, P. H., and Smart, I. H. (1984). "The Shape of Birds' Eggs." *Journal of Theoretical Biology* 106, 239–243.

Tricker, R. A. R. (1964). *Bores, Breakers, Waves and Wakes.* M. & B. Elsevier, New York.

Tricker, R. A. R. (1970). *Introduction to Meteorological Optics.* M. & B. Elsevier, New York.

Vergara, W. C. (1959). *Mathematics in Everyday Things.* Harper, New York.

Weinstein, L., and Adam, J. A. (2008). *Guesstimation: Solving the World's Problems on the Back of a Cocktail Napkin.* Princeton University Press, Princeton, NJ.

Whitaker, R. J. (1974). "Physics of the Rainbow." *The Physics Teacher* 12: 283–286.

White, P. R. (1958). "A Tree Tumor of Unknown Origin." *Proceedings of the National Academy of Sciences* 44, 339–344.

Williams, J. (1997). *The Weather Book.* Vintage, New York.

Young, A. *http://mintaka.sdsu.edu/GF/index.html*

Index

About the author

John Adam teaches mathematics at Old Dominion University, and his general interest is in applied mathematics and mathematical modeling, particularly in biology and currently in atmospheric optics. He is author of the book *Mathematics in Nature: Modeling Patterns in the Natural World*, published in 2003 by Princeton University Press (the paperback edition appeared in 2006). He is also co-editor of the book *A Survey of Models for Tumor/Immune System Dynamics*, published by Birkhäuser in 1997. He enjoys walking (of course) and nature photography, and is a frequent contributor to the Earth Science Picture of the Day (EPOD: http://epod.usra.edu/). A selection of his photographs can be found on his home page (http://www.odu.edu/~jadam). A British citizen, Dr. Adam received his PhD in theoretical astrophysics from the University of London. He was an undergraduate during the Monty Python years, and has never fully recovered, despite having taught for more than 25 years at Old Dominion. He is married and has three grown children, each of whom was born in a different country in the United Kingdom, and two wonderful grandchildren, John Mark and Caroline Grace. Their grandfather has been a frequent speaker at local universities, colleges, high schools, and civic and community groups. In 2007 he was a recipient of the State Council of Higher Education of Virginia's Outstanding Faculty Award. In 2008, *Guesstimation: Solving the World's Problems on the Back of a Cocktail Napkin*, written by him and his co-author Larry Weinstein, was published by Princeton University Press.